Causes of Development

THE DEVELOPING BODY AND MIND

Series Editor: Professor George Butterworth, *Department of Psychology, University of Stirling.*
Designed for a broad readership in the English-speaking world, this major series represents the best of contemporary research and theory in the cognitive, social, abnormal and biological areas of development.

LIBRARY

Tel: 01244 375444   Ext: 3301

This book is to be returned on or before the
last date stamped below. Overdue charges
will be incurred by the late return of books.

Chester
A College of the
University of Liverpool

Causes of Development

Interdisciplinary Perspectives

Edited by

GEORGE BUTTERWORTH
and
PETER BRYANT

HARVESTER WHEATSHEAF
New York   London   Toronto   Sydney   Tokyo

First published 1990 by
Harvester Wheatsheaf,
66 Wood Lane End, Hemel Hempstead,
Hertfordshire, HP2 4RG
A division of
Simon & Schuster International Group

Printed in Great Britain by BPCC Wheatons Ltd, Exeter

British Library Cataloguing in Publication Data

Butterworth, George *1946–*
    Causes of development : interdisciplinary
    perspectives
    (The developing body and mind)
    I. Title II. Bryant, Peter III. Series
    155

ISBN 0–7450–0664–7

1 2 3 4 5   94 93 92 91 90

Contents

Contributors

Professor Peter Bryant, *Department of Experimental Psychology, University of Oxford, England.*

Professor George Butterworth, *Department of Psychology, University of Stirling, Scotland.*

Professor Susan Carey, *Department of Psychology, Massachusetts Institute of Technology, Cambridge, Massachusetts, USA.*

Professor James Chisholm, *Department of Applied Behavioral Sciences, University of California, Davis.*

Professor Brian Goodwin, *Department of Biology, Open University, England.*

Dr Paul Harris, *Department of Experimental Psychology, University of Oxford, England.*

Professor Brian Hopkins, *Faculty of Educational Sciences, Free University of Amsterdam, Holland.*

Professor Robert Hinde, *MRC Unit on the Development and Integration of Behaviour, University of Cambridge, England.*

Professor P. N. Johnson-Laird, *Laboratory of Experimental Psychology, University of Cambridge, England.*

Professor Ivana Markova, *Department of Psychology, University of Stirling, Scotland.*

Dr James Russell, *Laboratory of Experimental Psychology, University of Cambridge, England.*

Professor Peter Slater, Department of *Biology and Pre Clinical Medicine, University of St Andrews, Scotland.*

Preface

The question of causal explanation in developmental psychology is an important and neglected issue. Development has been studied from a variety of perspectives and in many disciplines, yet it is difficult to say whether the resulting explanations are merely descriptive. Do developmentalists just describe the phenomena under investigation or do their theories also explain what causes development to occur? How can the developmental scientist make causal inferences about development? Causal processes have been suggested at the genetic, the embryological, the evolutionary, the social and the cultural levels but there has been little attempt to systematize the relationship, if any, between different types of developmental explanations. In fact, the variety of explanations may even be an impediment to the further development of the discipline. There have been trenchant and well-founded criticisms of the fragmenting effect that reductionist explanations have upon developmental science (Oyama, 1985) and the time seems ripe to examine whether any common principles may be discerned.

The issue of causal explanation is tackled in this volume from various theoretical perspectives and from methodological points of view. Eminent scientists from cognitive developmental psychology, developmental biology, ethology, embryology, social psychology and computer science were invited to consider the question at a conference held at the University of Stirling in 1985. Each offered a partial insight on the issue from their own discipline. It was apparent that factors considered causal from any one perspective might be completely ignored within another and we concluded that any one type of explanation, particularly when applied to human development, is *necessarily* incomplete.

The chapters collected in this volume have either been rewritten since the Stirling conference or are specially written for the book. The aim of the book is to consider the extent to which

juxtaposition of disciplinary perspectives may offer the prospect of
a unified developmental science within a causal framework.

Theory and method in causal developmental explanation

Part one considers theoretical and methodological approaches to
causal explanation. Brian Hopkins and George Butterworth offer a
systematic approach to various kinds of causal explanation. The
discussion encompasses the spectrum from the biological bases to
questions of continuity and discontinuity in development.
Following Aristotle's distinctions it is suggested that material and
formal causes may be distinguished from efficient and final causes
of development. The biological basis of development (the material
cause) possesses a certain, limited temporal priority, and has a
causal status. Correlatively, the forms of the organism and its
constituent sub-systems (formal cause), distinguish the same
organism at different points in development, and thus contribute
to causal explanation, but again in a static way. Developmental
explanation, however, must explain change and progression. This
requires consideration of antecedent–consequent relations (efficient causes) and recurrent regularity in development. Teleological
explanation, which asserts that development may be goal directed
or adaptive (final cause) also needs to be considered.

Contemporary accounts of development, such as Piagetian
theory, or socio-biological theories, or ethology are deeply rooted in
biological and evolutionary thinking and they attempt to
incorporate all four types of cause. Even so, they differ profoundly
in the extent to which each type of explanation is emphasized. For
example, heated debates occur about the extent to which
development should be considered a programmatic function of
genetic information or whether this metaphor is fundamentally
mistaken because it loses sight of holistic relations by asserting a
type of pre-formationism. Hopkins and Butterworth consider a
range of causal explanations, including simple mechanical
determination, statistical (probabalistic) determination, transactional, interactional and dialectical determination. It is concluded
that a spectrum of types of cause is needed to account for all
influences on development. What a unified explanation should

have in common is a concern for the critical aspects of the organism–environment relationship at different times in development. Such a focus would enable developmental science to move beyond over-general, causally vacuous explanation.

Given the complexity of the causal question there are further difficulties which face empirical research. Even if a developmental change can be described how can it be known if earlier and later states of organization *are* actually related to each other? In Chapter 2, Peter Bryant suggests that almost all the progress in cognitive developmental psychology arises from a description of the changing forms of cognition at different ages (formal causes) without reference to the antecedent–consequent relations (efficient causes) that may determine development. The reason for this omission is that most research has employed cross-sectional methods (i.e. different children are studied at different ages rather than the same child being studied longitudinally). Thus, causal models are never subjected to explicit test; putative causes remain implicit in the theory and are not demonstrated by research. Furthermore, Peter Bryant argues that some of the causal principles classically invoked by cognitive developmentalists, such as "equilibration" in Piagetian theory, or "the zone of proximal development" in Vygotsky's social theory are vacuous. (Just as empty is the concept of "maturation" which Hopkins and Butterworth object to in their chapter because it neglects to specify the multiple feedback processes that make maturation seem automatic.)

Peter Bryant argues that one of the roots of the difficulty of causal explanation lies in characterizing the origins of development in terms of what the child cannot do. In describing origins in negative terms, rather than by giving a positive characterization of the original abilities of the child, antecedent–consequent relations cannot be established. One methodological issue is therefore to characterize the origins of behavior in positive, adaptive terms. Other methodological issues are also involved in establishing causal relations in development. A causal test of the relation between an original and a later state of organization would involve longitudinal and intervention studies. A longitudinal prediction would enable a test of possible connections across behaviors in developmental time, e.g. whether infant visual habituation rate is related to subsequent verbal IQ. However, longitudinal studies alone do not permit causal inferences, i.e. that habituation rate in infancy is the antecedent "cause" of verbal IQ in childhood. An intervention study would be necessary to make inferences about specific connections between antecedent and consequent variables in cases where such specific links exist. For example, differential practice of skill x at age n should result in earlier acquisition of skill y at age $n1$ if there is a specific developmental link between them. The search for causal specificity in the antecedent–consequent sense, need not exhaust the possible range of causal relations that could exist between earlier and later stages of development. However, where the domain of interacting influences can be sufficiently restricted, it is possible to arrive at a reasonably satisfactory causal explanation for particular developmental outcomes. Furthermore, Peter Bryant pinpoints the methodological prerequisites for establishing such connections and this has important educational and therapeutic implications.

Causes of development in biology and ethology

Part two is devoted to developmental explanation in biology and ethology. Brian Goodwin is particularly concerned with the origins of form in embryogenesis and evolution (formal causes). He argues that in the case of universal patterns of development, as in embryogenesis, it may be impossible to isolate adequate single factors. Mechanistic conceptions of causality are expressed in static terms whereas in biology, process is primary. Once it is accepted that change is the fundamental quality of living systems then the task is not to explain what *causes change* in the sense of a force externally applied to a static organism but to account for dynamic stability in all its forms. Goodwin's insight is perhaps the most revolutionary for it demands a dynamic foundation for developmental science and it constantly refers back to the whole organism, in context, as the appropriate unit for developmental analysis.

Goodwin's theory certainly meets the holistic requirements of contemporary critics of reductionism in developmental explanation. Oyama (1985), for example, suggests that development in ontogeny occurs as a result of the functioning of nested causal systems. Her causal explanation is couched in terms of interactions among factors that control the developmental process. Their order of emergence is a result of three levels of interaction:

1. Control of development arises through mutually selective interactions among constitutent processes.
2. Control of development arises through emergence in hierarchical levels in the sense that processes at one level interact to give rise to new processes at the next (and later levels may in turn be reflected in earlier ones).
3. Control of development emerges through time and is sometimes transferred from one process to another.

From this dialectical perspective the essential element of a causal account is the reconstitution of the whole organism from its parts, conceived as mutually interacting control systems. The genes or any other sub-system should never carry the whole burden of explanation.

Peter Slater considers the question of causal explanation from an ethological perspective. He suggests that causal explanation within ethology and socio-biology has been much misunderstood. Two types of cause may be distinguished: proximate causes concerned with the functional (survival) value of behavior and ultimate causes concerned with the evolutionary history of the species. Both types of causal explanation hinge crucially on natural variation in gene frequencies. Since all behavior in some sense has a genetic component, behavior–genetic explanation offers a causal theory of that variability. However, this does not mean that genes determine behavior. Genetics affects the relative frequency of behavior, it contributes to the variability but does not cause it mechanistically.

Most of Peter Slater's chapter is devoted to a review of the interacting environmental and species typical factors giving rise to variations in the development of bird song. He discusses many interacting factors considered as proximate causes by ethologists. These include sensitive periods for learning; the spatial structure of the environment of rearing; the presence of conspecifics and the possibility that "templates" for basic components of species typical songs may be inherited. An ultimate (final) cause for the development of bird song may be its critical role in mate identification and reproduction. This explanation in turn returns to behavior-genetics and the causal hypothesis of the maximization of inclusive fitness.

Causes of cognitive development

Part three considers one of the key areas of contemporary developmental psychology: the problem of causal explanation in cognitive development. Cognitive developmental theory is concerned with how children learn to reason; how they become able to derive new premises from their existing knowledge. Phillip Johnson-Laird considers what might cause the development of the ability to make deductions. He argues that there is an innate, evolutionary basis for deductive reasoning which enables valid inferences to be distinguished from invalid ones. The cause of development does not lie in the ability to make deductions, he suggests, but in acquiring the ability to imagine mental models which enable the child to refute invalid deductions. The other essential developments are in language and memory. These developments enable the child to master the syntax and semantics of logical terms and to hold in mind several alternatives. A feature of cognitive development is the recursion of mental models on themselves, so that the mind forms a model of its own abilities on the basis of its innate machinery. Formal logic is not an original characteristic of the mind, it constitutes a systematization of the procedures for seeking counter examples and it is learned under tutelage.

Theories of cognitive development concern a restricted domain yet, however circumscribed, causal explanation seems to draw the scientist into a wider and wider circle of antecedents recognized as relevant. James Russell inverts some of the arguments in preceding chapters to suggest that holism may actually be a problem in causal explanation since it implies failure to set manageable limits on a domain. To be practical, causal explanation should specify the particular conjunction of *unnecessary but sufficient* explanations required for any cognitive skill. Under some circumstances a simple description may specify a sufficient cause; it all depends what purpose the explanation is intended to serve.

Susan Carey adds another note of caution, which is that we should do our best to be sure of what it is that develops before we plunge into theories about the causes of that development. She uses the example of the question whether cognitive development consists of the acquisition of underlying general skills, which affect behavior in several domains, or alternatively the acquisition is of specific knowledge about specific domains. These different possibilities demand quite different kinds of causal hypotheses.

Causes of social development

Part four moves towards an examination of inter-personal processes and their causal role in development. Robert Hinde argues that social development is a dialectical process, with cause and consequence inextricably interwoven. He argues that both organism and environment are continually active and changing; the most important feature of the environment being other persons. A dialectical explanation is similar to a process model of development, in so far as activity and change are primary, while the "static" elements of the system are secondary and derived. On this view, it is not possible to isolate factors which mechanistically push development along, not least because static "snapshots" of the system are only momentary states of affairs.

One advantage of dialectical explanation is that it captures simultaneously, different levels of the developing system. Robert Hinde suggest that development at the individual level is affected by higher levels of complexity in at least three ways: (1) for survival, e.g. the baby must be fed and protected; (2) for socialization, e.g. the child must learn from others appropriate social behaviors; and (3) for acculturation, e.g. the child must learn the belief systems current in the particular culture. The child is formed by and forms part of a network of relationships that are causally implicated in development. The simultaneous influence of different levels of complexity also allows explanations for development which extend beyond the limited teleology of the child. Parents may guide development according to long-term culturally determined goals which enter into the causal nexus. Robert Hinde also gives extensive consideration to the possibility that the influence of biological factors on development may be explained dialectically. For example, hormonal influences on sex differences in aggression may be modified by the social conditions of rearing male and female rhesus monkeys. A causal explanation which incorporates biological factors need not imply biological determinism.

This dialectical approach to social development is further pursued in the chapter by Ivana Markova. She suggests that three essential criteria define human social development: co-development of organism and environment; progressive differentiation and hierarchical integration of the structure of the organism; and conscious awareness of social others.

A dialectical approach captures the relational nature of these terms; they are dyadic relations which cannot be properly understood in isolation. The intelligibility of the term "organism" is dependent on its counterpart "environment" and they both co-develop; just as a man becomes a father at the moment his child is born or the child a son or daughter by virtue of the relationship with the parent.

The emphasis on the dyadic nature of social development is intended to redress prevailing asymmetric theories of social influence. Ivana Markova considers the dialectical explanation of social development in the light of the theories of James Mark Baldwin and George Herbert Mead, with particular reference to the causal role of communication.

Culture and causes of development

Given a common humanity and a shared biological heritage, how does diversity of culture put its particular stamp upon development? The final part of the book extends the examination of causal questions to the cultural level. Paul Harris points out that diverse cultures foster common properties of cognitive development as far as universal physical phenomena are concerned. Objects are permanent and water is conserved no matter which culture and this makes a cultural influence hard to discern. However, some psychological processes, notably emotional expression, show great cultural diversity. This in turn may lead the child to a culturally dependent meta-theory of mind. This research extends current interest in meta-cognition (Astington, Harris and Olson, 1988) to the child's own conscious awareness of the operations of mind.

An example of cultural influence on cognitive development comes from Margaret Mead's description of the acquisition of magical beliefs by Manus children. Young Manus share the same concept of cause and effect as Western children; a canoe goes adrift because it is not moored. Adult Manus, however, attribute evil intentions to the canoe; they impose a culturally specific meta-theory on the event. Paul Harris argues that the child may acquire many culturally specific meta-theories.

One version of a theory of emotional development would simply assert that culturally specific conditions, e.g. a child losing the cattle in a herding culture or losing a toy in a Western culture may both elicit sadness. However, a second, more complex explanation may apply. For example, shyness may be valued in some cultures and may be experienced quite differently by the child when it elicits approval, rather than reproval, from adults. In that case, the culture shifts the balance among universal emotions, placing them in conjunction or disjunction and thus exercises a specific causal effect. A third possibility is that the culture constitutes the emotion: this dialectic presupposes that knowledge of a particular set of cultural assumptions must be acquired before the emotion (e.g. machismo) can be experienced. Paul Harris argues that while the core of mental development may be universal, the evidence shows that by adolescence, culturally specific meta-theories are imposed on a wide range of mental phenomena.

The final chapter, by James Chisholm, an anthropologist, demonstrates one way in which biological and cultural explanations of development may be integrated. He argues that socio-biological causal models have been mistaken in their emphasis on the genotype. The socio-biological approach needs to be modified since social and cultural factors, which influence reproductive success, operate at the phenotypic level. The success of the phenotype determines which genes are transmitted to the next generation. A single genotype is capable of giving rise to a wide variety of environment-dependent phenotypes. Individual development gives rise to the particular phenotype. Hence, what is required is a co-evolutionary theory of human biology and culture, itself based on a theory of development.

The geographical range of humans and the wide variety of subsistence systems present the developing young with an unpredictable variety of tasks. The prevailing economic conditions may determine patterns of upbringing, whether by parents or peers, perhaps with father absent. The effects of culture differ systematically between subsistence and other societies. Thus, once the child acquires independent locomotion much early experience in subsistence cultures may be obtained from peer groups and this in turn, may have long term adaptive consequences. James Chisholm suggests that where access to resources depends on acquiring skills within the peer group, long term socialization may be toward skills that ensure maximum reproductive effort even when few offspring survive. In societies where early care is parental, the pattern of socialization may encourage high investment in relatively small numbers of children.

Other cultural influences that were considered at the Stirling meeting, but which have had to be omitted from consideration here, concern the role of scientific culture on development. Richard Gregory spoke eloquently on the possibility that the products of scientific enquiry are not only derived from but also act as amplifiers of human cognitive capacities. The material culture as constructed by humans, constitutes a kind of external representation of the mind and of human intentions. It is not easy to explain how each succeeding generation so readily masters the advanced technology of the parent generation. A theory of the contribution of technology to development would also be part of a comprehensive, causal theory of culture and cognitive growth (see Richard Gregory, 1981; Susan Carey, 1985; and Chris Sinha, 1988).

Conclusion

When we set out to examine the question of causal explanation in developmental psychology it was our impression that the field was diverse and that explanations were mainly non-causal. The deliberate juxtaposition of biological, social, cognitive and anthropological perspectives herein suggests that there does exist a rather more integrated relation between these levels of explanation than we had suspected.

Different types of causal explanation seem suited to different purposes. These explanations range from antecedent–consequent relations (not necessarily conceived as a force externally applied to the organism) to accounts of the dialectical relations among multiple constituent systems. The former type of explanation may have particular force where a domain is relatively circumscribed. Indeed, one problem in arriving at useful explanation is to restrict the range of causal influences to the sufficient ones (without considering all the necessary conditions).

The mutual dependence of evolution and development in many of our attempts at causal explanation is also apparent. This is unlikely to have been an accidental focus of this particular group of contributors. Rather, it reflects the fact that theorizing about causes within post-Darwinian developmental psychology will inevitably introduce an evolutionary dimension. What causes development is intimately linked to what causes evolution and this is the crux of the problem of explanation. While much has been omitted from our discussions what has been said holds the seeds and perhaps the promise, of a unified developmental science couched in causal terms.

George Butterworth and Peter Bryant
University of Stirling and University of Oxford. January 1989.

Acknowledgments

We are grateful to the Developmental Psychology Section of the British Psychological Society and to the Economic and Social Research Council of Great Britain for financial support.

References

Astington, J. W., Harris, P. L. and Olson, D. (1988), *Developing Theories of Mind*, Cambridge: Cambridge University Press.

Carey, S. (1985), *Conceptual Change in Childhood*, Cambridge, Mass.: Bradford Books.

Gregory, R. (1981), *Mind in Science*, Harmondsworth: Penguin Books.

Oyama, S. (1985, *The Ontogeny of Information*, Cambridge: Cambridge University Press.

Sinha, C. (1988), *Language and Representation*, Hemel Hempstead: Harvester Wheatsheaf.

Part I   Methodological and Theoretical Approaches

1 Concepts of causality in explanations of development

Brian Hopkins and George Butterworth

Aristotle suggested that any scientific enquiry requires answers to four basic questions: what matter was involved, what was responsible for the presence of the object of investigation, what forms did it possess, and what was it striving for? Answers to these questions have been traditionally termed Aristotle's four causes: "material" (that *from* which something is produced); "efficient" (that *by* which something is initiated); "formal" (that according *to* which something is done) and "final" (the sake *for* which something is done).

At the root of these questions lies a more basic concern with what constitutes an individual. For Aristotle, individuals have potentialities for change, for *becoming* and not just for *being*. The realization of potential depends on a form-giving principle (the eidos) being directed towards some end state (the teleos). In proposing formal and teleological explanations of change, Aristotle laid the foundation for the concept of epigenesis which

Table 1.1: The epigenetic approaches of Waddington and Piaget involve the four types of Aristotelian causation (material, formal, efficient and final). Each of these types is identified with a particular aspect of these two complementary approaches to development.

	Material	Formal	Efficient	Final	Causes
Waddington	Sets of cells and extracellular substance	Chreod	Induction	Homeorhesis	
Piaget	Genetic and maturational	Structure d'ensemble	Physical and social environment	Equilibration	

lies at the heart of current theories of development. He emphasized that the necessary and sufficient conditions of organic change could only be determined by taking all four causes into consideration.

Epigenetic explanations of development, such as those of Waddington (1957) and Piaget (1971), attempt to take into account material, efficient, final and formal causes (see Table 1.1). Such materialistic explanations treat final cause (or goal-directedness) as the product of complex interactions between the other three sorts of causes (see Figure 1.1) and involve (among other things) programs and (negative) feedback loops.[1]

Delbruck (1971) and Mayr (1982) argue that Aristotle's "eidos" or formative principle was one of the greatest conceptual innovations in biology. In the language of modern biology it is

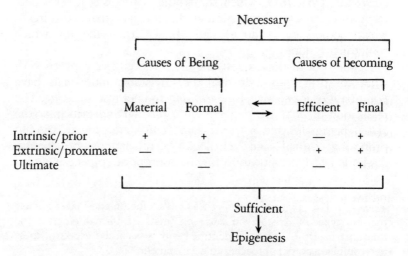

Figure 1.1: Epigenetic explanations of development involve the four types of causation. Material and formal (intrinsic, prior causes of being) together with efficient (extrinsic, proximate cause of becoming) are necessary but not sufficient to explain epigenesis. For sufficiency, final cause is needed which is the outcome of interactions between formal and efficient causes (i.e. between intrinsic and extrinsic causes) which result in an ultimate cause.

+ = applicable to one or more of the types of causation
− = not applicable

compatible with the concept of the ontogenetic program, a direct descendant of Aristotle's formal cause. However, it is important to bear in mind that Aristotle's epigenesis was essentially vitalistic and that his formal cause only received a materialistic explanation acceptable to embryologists late in the nineteenth century. The modern concept of the ontogenetic program rests on three main assumptions:

1. The program contains all instructions necessary for development to proceed from the cellular level to that of the reproductive adult.
2. The instructions are not programmed in detail. They set limits on environmental action while at the same time providing for competencies to react.
3. The structure of the program's instructions cannot be modified directly by the environment. The environment does not instruct but rather selects and exploits the competencies determined by the program.

Returning to Figure 1.1, a distinction can be drawn between causes of *being* (material and formal) and causes of *becoming* (efficient and final). In the history of science, this dichotomy has been maintained in contrasting ways. In some models of change, material and efficient (proximate) causes were emphasized while in others priority was given to formal and final (ultimate) causes. These various models will be considered in terms of their strengths and weaknesses. But before attempting this, it is necessary to discuss what is meant by causality.

On causality and determinism

Kagan *et al.* (1978) consider the pervasive assumption that: "The interactions between biological and environmental forces determine the psychological growth of the organism" (Kagan *et al.*, 1978, p. 44). They ask: "What in heaven's name does that fourteen-word sentence mean? (p. 44)." The meaning depends on the particular concept of interaction for this can be quite different from case to case. In Bunge's (1959) view interaction or reciprocal causation is one among several categories of general determination. Terms such as causation, determination, causality and determinism are often not clearly separated. Bunge makes a convincing argument for treating causal determination as a special form of general determination, because in modern science many noncausal categories are employed such as structural and dialectical determination though they are often erroneously cast in causal language. He points out that causality has three distinct meanings: causation (the general causal relation between two things when the first is necessary and/or sufficient for the occurrence of the second), the causality principle (a statement of the law of causation in the form "The same cause always or invariably produces the same effect"), and causal determinism or causalism (the doctrinal assertion of the validity of the causal principle).

According to the causal principle cause and effect are tied together in a necessarily constant and unique way such that the "tying together" is brought about by external conditions (e.g. maternal separation invariably causes distress in infants). In contrast general determination does not involve constant and unique conditions; neither is it restricted to external agencies nor to quantitative variations in which qualities are fixed (as in mechanical determination). The principle of determinancy is the hypothesis that processes with definite quantitative and/or qualitative characteristics take place in one or more determinate ways of becoming. These are lawful and acquire their characteristics by processes which evolve from preexisting conditions. Thus, general determination allows for causal, mechanical, statistical, teleological and other forms of determination.

Bunge (1959) proposes a hierarchy of categories of determination ordered according to their increasing complexity in which each level is characterized by a particular newness of its own but at the same time being rooted in the lower levels (see Table 1.2). These are irreducible but interconnected so that no type of determination operates in exclusion of all others. This hierarchy offers a framework within this chapter for outlining the various approaches to explanation in contemporary developmental psychology. The discussion will be illustrated by reference to the major theories of development and to various well-known research programs. Teleological determination (Aristotle's final cause), mechanical and statistical determination (efficient or material

Table 1.2: Bunge's (1959) spectrum of categories of determination hierarchically arranged.

Category	Description	Basic question(s)	Directionality
Dialectical determination or qualitative self-determination	Of whole process by internal disequilibrium and eventual synthesis of opposites	How When	Bidirectional (dynamic)
Teleological determination	Of means by ends or goals	Why Where to	Uni- or bidirectional
Structural or holistic determination	Of parts by whole and *vice versa*	How	Uni- or bidirectional
Statistical determination	Of end result by joint action of independent or quasi-independent entities	How much	Unidirectional
Mechanical determination	Of consequent by antecedent with addition of efficient causes and mututal actions	How much	Unidirectional
Interaction of reciprocal causation	Of consequent by mutual action	How	Bidirectional (static)
Causal determination or causality	Of effect by external (efficient) cause	Which one	Unidirectional
Quantitative self-determination	Of consequent by antecendents	How much	Unidirectional

causes), and structural determination (formal cause) are discussed as the major examples of causal theory in contemporary developmental science together with more recent ideas about interactionism and dialectual determination.

Teleological determination

Teleological determination takes two forms: dynamic teleology and static teleology. Static teleology means that a particular characteristic or structure of the organism is assumed useful for a certain purpose. In Aristotelian terms this is formal causality. For example, predatory animals have eyes in the front of the head, which presumably helps them in the kill, while the prey have eyes on the side of the head, which facilitates escape. Another example concerns the allometric shape changes, particularly of the head and face, which occur during growth. During early infancy the head and face assume a characteristic infant shape (large head, bulging forehead, flat face, and large eyes relative to the face) as do the limbs (short and pudgy). According to Lorenz (1965) these features, combined with small size and clumsy behavior, serve the purpose of innate releasing mechanisms for parental behavior.

Dynamic teleology, which relates to Aristotle's idea of final cause, involves three parallel interpretations. Firstly, it can be depicted as foresight of some particular goal which is assumed to be already in thought and so serves to guide present action. Secondly, and most commonly, it involves explanations of behavior in functional terms where questions of the form "what is the adaptive value of this behavior?" are the focus of interest. Thirdly, it posits a preprogrammed directiveness to development in which species-characteristic end states can be achieved through a variety of means. These various interpretations of dynamic teleology share a concern for explaining where the organism is going rather than where it has come from.

Purposive explanation has been a primary debating ground for psychologists ever since McDougall (1932) insisted that human behavior has to be analyzed with reference to ends sought, and not simply in mechanistic terms.[2] In McDougall's view purpose is a defining characteristic of human behavior which can be studied scientifically and by ignoring it psychology omits a basic form of theoretical analysis. In linking purpose with consciousness, McDougall drew a dividing line between teleology in psychology and biology. Functional and programmed interpretations are biological and therefore, according to McDougall weaker than his purposive use of teleology (Boden, 1978).

In developmental psychology, teleological determination has been most forcibly expressed in Werner's (1948; 1957) principle of orthogenesis as applied to perceptual, motor and conceptual systems. His orthogenetic principle states: "that wherever development occurs it proceeds from a state of relative globality and lack of differentiation to a state of increasing differentiation, articulation and hierarchic integration" (Werner, 1957, p. 126).

The dynamic teleology of Werner's orthogenetic principle is formally specified by four rules. In the functional analysis of development, progression is from syncretic to discrete, i.e. from relatively undifferentiated or fused to differentiated, such that there is increased individuation of systems of action. Structural analysis involves progression from diffuse to articulate, from lack of structural differentiation to clear-cut boundaries between integrated systems of action. The increasing adaptiveness of the organism to endogenous and exogenous changes is explained by two complementary sets of formal rules: rigid to flexible and labile to stable development. Undifferentiated levels of development show relatively rigid but unstable modes of organization in which the organism cannot adapt responses to marked changes coming from within or without. At later levels lability and stability coexist so that small endogenous or exogenous variations can push an organism out of its developmental course. This in turn activates self-stabilizing mechanisms to restore the directiveness of development.

Piaget showed a more reserved commitment to teleological determination. He rejected vitalistic or finalistic interpretations of teleology based on the preevolutionary doctrine which held that

the activity of the organism or of the intelligence is strictly limited to the utilization of the environment according to some pre-established plan or to the contemplation of it as intellection. (Piaget, 1971, p. 103)

Teleology in this sense amounts to attributing some preestablished harmony between organism and environment without explaining how this arises. Piaget (1971) stressed the formal parallelism between biological and cognitive autoregulation. He proposed teleonomy, a reinterpretation of teleology in cybernetic terms, as a means by which this question can be answered. Teleology for Piaget (1971) can be interpreted as

processes that were both set in a certain direction and capable of self-direction, sometimes able to anticipate what was coming and playing a useful part in an organized system so that they could be summed up as what are recognized by common consent to be finalized systems. (p. 132)

Examples of the application of teleological concepts to early development can be found in Bowlby's (1969) theory of attachment. An interesting case is Konner's (1972) research dealing with the neurological reflex responses of a small sample of !Kung infants. These infants are carried upright in a sling which maintains skin-to-skin contact with the mother. In this situation a number of responses occur in the infant to maternal postural changes. The !Kung baby shows not only head lift but also placing responses and stepping movement (see Prechtl, 1977). Konner speculates that these elicited movements may function to lessen the burden on the mother to make postural readjustments in her infant and may prevent smothering in her skin. Another example concerns the palmar grasp which can be facilitated by sucking. Konner notes that the babies grasped and clung to the mother's clothing and beads as well as to the skin of the breasts and concludes that the response serves to stabilize the hold of the infant's mouth on the breast. Again, this reduces the need for constant vigilance by the mother.

The debate over whether teleological determination is a relevant explanatory framework for the developmental sciences has generated more heat than light. A main objection has been that teleological interpretations can be restated in terms of causal analysis, making certain assumptions about the nature of causal explanation (Nagel, 1961). In order to make this translation it is necessary to assume that functions, goals and effects are synonomous and to classify effects as necessary conditions and causes as sufficient conditions. In some cases this can be achieved but in others it is patently illogical. While all functions are effects, not all effects are functions. For example, an infant crying may be the effect of hunger but crying is not a function of hunger. Another objection relates to the problem of goal failure. An infant cries but mother does not hear or decides not to respond. Had she responded by calling to or approaching her infant then a teleological interpretation of crying would be permissible but given the fact she made no observable response one is forced to abandon this form of interpretation. The fallacy of this objection lies in assuming that a teleological system has to be perfect all of the time which in fact is hardly ever the case. Even the most perfect of teleological systems sometimes fails to reach a goal but this does not make it any the less teleological.

Teleological systems can achieve the same end goal through alternative pathways. They are near the open end of the systems continuum and demonstrate equifinality. Such functional equivalence forms the very basis for adaptation in ontogeny and phylogeny. It is so immanent in nature that the scientific status of teleological determination can never be satisfactorily replaced by mechanistic interpretations.

Contemporary biology distinguishes teleological from nonteleological systems on the basis of a number of criteria drawn from embryology, cybernetics and evolutionary biology which do not have recourse to McDougall's notion of conscious purpose. These are: a high incidence of preferred states, negative feedback and/or programs, and the derivation of such systems by selection processes. By themselves none of the criteria is successful but taken together they are difficult to apply to non-teleological systems. In embryology preferred states are treated as information stored in stable DNA molecules. This stored information then serves as the reference standard or goal state to which the system is oriented. For a system to be considered teleological, it is necessary that the goal state occurs with a higher incidence than other possible states of the system. Deviations from a reference standard can be corrected either by negative feedback loops or by programs (codes of information) which provide alternative pathways similar in nature to the original pathway. (e.g. Mayr, 1974).

A historical criterion of selection processes is needed to account for the way in which the information which guides development has been coded into the genome. At the very least it should be acknowledged that evolutionary processes have imposed constraints on the availability of functional equivalents by which a species attains characteristic preferred states in development. Without a historical perspective a teleological explanation of development is simply linear causality made circular in nature (Overton, 1975) such that *causes cause causes to cause causes* to use Wilden's (1972) alliteration.

The modern idea of a genetic program dates back to Weismann (1893) and his notion of a generative germ plasm. At last, it seemed, a materialistic basis had been identified for Aristotle's notion of final cause, thereby resolving the epigenesis–preformationism debate. Subsequently Weismann's notion was metaphorically interpreted as a (genetic) blueprint (Løvtrup, 1974): his immortal germ plasm was treated as if it contained a detailed plan or "anlage" of how the organism should be constructed. Development consisted in following the step-like instruction mapped out in this plan in analogous fashion to a master builder constructing a house. This neo-preformationism proved to be biologically implausible because among other things, it would exclude the possibility of embryogenetic modifications acting as a source of phlyogenetic change. Some other metaphorical source had to sustain the belief that development followed precoded instructions or a directive agency. This was found in the theory of automata (Longuet-Higgins, 1969). Now a genetic program was treated as if it were an algorithm (i.e. a procedure for carrying out a complicated operation or instruction by means of a precisely determined sequence of simpler ones). Like an algorithm, it was taken to have an *alphabet* (sequence of purine and pyrimidine base pairing in nucleic acid), *a vocabulary* (the set of triplet codons that code for a particular amino acid) and a process of *compilation* (protein synthesis). With this analogy, a program was seen as regulating the outline of development while the process of epigenesis sketched in the major details.

Having reduced development to a process of molecular differentiation analogous to the operations performed by a computer, certain questions and problems still remained unanswered (Longuet-Higgins, 1969); how did existing programs originate (the origins of life problem)?; how have they evolved (the process of evolution problem)?; and how are they implemented and operationalized (the process of epigenesis problem)? In short, how such a program can generate adapted forms during evolution and development has remained an unresolved problem. This has led to the recent opinion that it is unlikely that development is a programmatic phenomenon. According to Stent (1981; 1984) the whole notion of a genetic program is rooted in semantic confusion. For development to be considered programmatic, there must be an isomorphic correspondence between it and the program just as a computer program has a structure isomophic with the sequence of operations carried out by the hardware. On this criterion, little about development is programmatic. Rather, development is a regular but non-programmatic phenomenon which follows natural laws and not a predefined script somehow encoded in the genes. Its regularity results from a cascade of contextually constrained stochastic interactions in which one thing simply leads to another (see Wicken, 1981). In contemporary embryology, this structuralist view is reflected in models of developmental transformation which ignore genetic programs, and even genes themselves (e.g. Newman and Leonard, 1983). Here the organism is equated with a field whose solutions constrain the potential number of forms an organism can attain. Accordingly, development can be portrayed as a sequence of field solutions mainly emerging from processes intrinsic to the field. It is beyond the scope of this chapter to discuss these models which signal a break with neo-Darwinistic concepts of development. The reader is referred to Goodwin's chapter (in this volume) for further information on this natural-physical approach to development.

Mechanistic determination

Mechanical determination in the form of antecedent–consequent relationships is a widely accepted interpretation of developmental processes. Antecedents are typically couched in terms of past reinforcement history while efficient causes constitute current reinforcement. Most often antecedent determinants of behavioral consequents have been given in environmental terms and to a lesser degree as biological factors. Whatever the precursor, this "push-model" of quantitative change supports notions of psychological continuity. Psychoanalytical theory stresses the overriding importance of the role of parents during the early years of development in determining the course and outcomes of basic psychic structures.

The biologist interprets epigenesis as the formation of a reproductively mature organism by a stepwise progression in which the emergence of later steps depends on the successful completion of earlier steps. This interpretation has had a profound influence on constructivist theories of individual development. At the psychological level, Piaget's explanation of development as a process of *being* maintained in the face of *becoming* has of necessity incorporated a form of mechanical determination. In Piagetian theory, psychological structures acquired during one period of development are subsumed and deterministically integrated into those of the following period. Furthermore just as with psychoanalytic theory Piaget's constructivist perspective has a strong allegiance to the evolutionary doctrine of recapitulation which posits continuity across phyletic levels.

The study of human development abounds with so-called developmental laws which typically take the form of antecedent–consequent relations: "If x has the property P at time t, then x has the property Q at time t' later than t" (Nagel, 1961, p. 76). An example of such a law is: "activity always precedes reactivity in the development of the nervous system". Such laws are not explanations but descriptions of a relatively invariant sequence of functional changes. It is notoriously difficult to explain how (and why) antecedent and subsequent functions in a developmental sequence are interconnected. Yet the assumption that successive events in ontogeny are somehow causally related remains strong. What is ignored is the possibility that some structures and functions may be transient. They are adaptive only for a restricted phase of development and may be unnecessary for (and even incompatible with) adaptation at later phases. Oppenheim (1981) and Prechtl (1981) discuss a number of such transient phenomena which occur during the development of non-metamorphic vertebrates. What is required in the study of development, is a theory of transition which can account for such phenomena while at the same time explaining how earlier functions are transformed into more mature abilities.

Mechanical determination as outlined by Bunge (1959) concerns more than just antecedent-consequent relations. It includes efficient causes and mutual actions (i.e. reciprocal causation). An example of the use of the comprehensive form of mechanical determination can be found in Klaus and Kennell (1976) on mother–infant bonding (see Figure 1.2). Here, particular antecedent events combine with certain efficient causes in determining the quality of first contact, which in turn has consequences for the social and emotional development of the child. In such a model, outcome is more or less fully determined by a set of complex initial conditions. No room is given for qualitative self-determination through intrinsic regulating mechanisms of the developing organism. Change can only come about by altering external circumstances on this mechanical model.

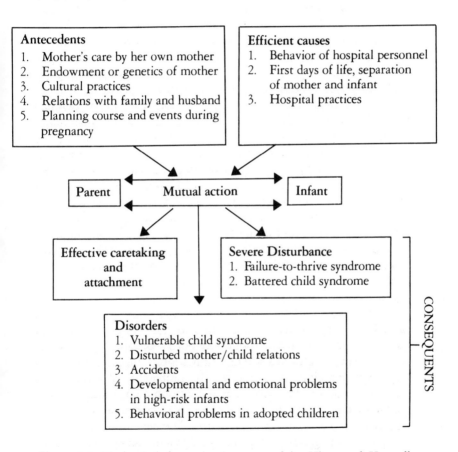

Figure 1.2: Mechanical determination as used by Klaus and Kennell (1976) to explain disturbances in development outcomes. (Adapted from Klaus and Kennell, 1976, Figs 1–4)

Statistical determination

Under the guidance of quantum mechanics, predictability is nowadays couched in terms of probabilistic relations to such an extent that no science asserts an invariable occurrence of relationships. In a sufficiently long series of trials, one event's occurrence is followed by another with a relative frequency or probability which in itself is invariable. The essence of statistical determinacy is an epistemological probabilism, which, looked at another way, attempts to account for that peculiar, and often unrecognized, form of determination called chance. It is the doctrine which reminds us, particularly in the behavioral sciences, that only probably general propositions can be asserted about empirical matters.

When physicists discovered that the causality principle was less than comprehensive in explaining the action of atomic particles, it marked the demise of causal determinism and with it any beliefs in linear causality. Acceptance of Heisenberg's uncertainty principle, Schrodinger's wave mechanics and Boltzman's laws of thermodynamics led to the replacement of simple causality with notions such as organized complexity in which cause and effect were difficult to detect: causal determinism was now replaced by probabilistic determinism. Instead of assuming unidirectional causality as in classical mechanics, reciprocal causality was deemed to be more relevant for the problem of explanation. In Aristotelian terms, material and efficient causes reciprocally interact in a probabilistic way leading to the emergence of formal causes or organized complexity.

Both Bunge (1959) and Nagel (1961) point out that a probabilistic perspective exludes rigid causality as put forward by the causal principle but without a loss of predictability which can be assessed through statistical determination.[3] Thus, in a matrix of transition probabilities, p assesses the probability of a transition from behavior 1 to behavior 2, as one among many other possible transitions. In a large number of cases (N), the transition $1 \rightarrow 2$ is almost certain to occur about Np times. The next step is to account for the determinants of the probabilities in question. In some instances suitable assumptions can be made about the statistical distribution of initial conditions permitting the application of causal laws, thus moving one closer to a causal explanation of the phenomena under study. The assumption here is that statistical determination probably arises from processes situated in other categories of determination at deeper levels in the hierarchy. The problem is that many statistical laws (e.g. in physics) still exist for which no causal explanations have been found.

Salmon (1975) has been a strong advocate of statistical determination as a causal explanation of natural processes. In his model an explanation is treated as an assemblage of factors that are statistically relevant to the occurrence of the event to be explained, accompanied by an associated probability distribution. An example is the preference of newborns to turn their heads to the right (Michel, 1981). Factors assumed to be relevant to this behavior include: head shape, sex, birth position and parental handedness. Newborns are then subdivided into a series of subclasses in terms of such factors. For each subclass in the resulting partition, the probability of right "headedness" for members of that subclass would be determined. If any factor did not make some difference in the probability of right "headedness" it would be discarded as irrelevant (i.e. as lacking in explanatory value). Thus, in probabilistic explanations it is important to find appropriate reference classes to account for a given probability distribution.

The apparent bond between statistical determination and causality can lead to the two terms being used interchangeably as Salmon (1975) appears to do. It is an attempt to reduce causation to a statistical concept in the sense that if events of class A reliably follow events of class B then a causal relationship is indicated. But this is a mistake since what is really indicated is not a causal connection but the replicability of a finding to some level of probabilistic confidence. What brings about the connection between the two sets of events has to be sought in other forms of explanation.

Both statistical and mechanical determination treat development as a genesis without structure. A quite different viewpoint arises when teleological determination is integrated with fieldbased approaches to structuralism, namely, a genesis with structure, an integration that characterizes Piaget's vision of development.

Structural determination

Kronenfeld and Decker (1979) argue that Piaget's structural approach offers one of the best insights into how thought evolves in the individual. But what is this structuralism to which Piaget has made a unique contribution? Definitions abound, but the conveyed meaning, as Overton (1975) points out, "... teeters on the brink of vacuousness (p. 63)". If it is taken as a content-free approach, concerned with going beyond the observational level by means of constructing models assumed to represent underlying organization, then just about every researcher in every field of scientific endeavor can claim to be a structuralist of some sort. In Overton's (1975) view structuralism is a paradigm concerned with the nature of structures that is inevitably couched either in organismic or mechanistic terms, a contrast between holistic and elementaristic structuralism.

Elementaristic structuralism treats structure as a derived construct which can be explained by other more basic constructs. Here structure is nothing more than the elements of the object and their relations, an approach which is common to Guilford's (1967) psychometric analysis of intelligence and Gagné's (1968) learning theory. In contrast, holistic structuralism, as represented by Piaget, regards structure as a primitive construct with a necessary role in explaining and understanding objects and events. The whole or structure (which Overton equates with organization) is necessarily prior to its parts and one should not begin by decomposing it into these parts. Unlike elementaristic structuralism, this doctrine views antecedent events and efficient causes as having an important but limited explanatory role. They can explain rate and ultimate level of development or momentary behavior when put forward in terms of hereditary or training programs but not its ordering. In this sense structures and their laws of transformation are similar to Aristotle's final and formal causes.

In developmental biology, criticism of the genetic program concept has led to alternative theoretical considerations which have their roots in the holistic structuralism of field physics. The field concept has had wide currency in embryology, particularly in depicting regional changes in spatial relationships between cells and tissues, but it too has been a target for criticism. For example, Child (1941) considered it to be only a superficial explanation; little more than a verbal refuge for what is in effect poorly understood. In an attempt to reinstate the theoretical significance of the concept, Waddington (1957) proposed that a developmental field could be characterized by a self-stabilizing pathway or chreod: a spatial-temporal system in which the controlling influence of particular alleles are in part modified by environmental circumstances. More recently, Thom's (1975) catastrophe theory has increased the relevance of the field concept for the developmental sciences. In such field theories, development is depicted as a self-defined sequence of field solutions, each of which set boundary values for successive solutions in a self-organizing mode of operation. While fields are stabilized or constrained by the products of genetic activity, their solutions are generated by processes intrinsic to the field. An excellent introduction to the application of the modern field concept in developmental biology is given by Goodwin (1985; see also his chapter in this volume) who also shows its relevance for Piaget's structuralism. Further insights can be gained from Kugler (1986) who uses field equations and generative rules for explaining developmental transformations in a theoretical perspective based on the structuralist approaches of Bernstein (1967) and Gibson (1979).

Overton (1975) discusses many more aspects of the metaphysical distinction between elementaristic and holistic structuralism, but in the final analysis his presentation does not progress beyond generalities. What is meant by the concept of structure? Does it have universal application or is it restricted within each of the various levels of analysis? To what extent is it feasible to draw structural isomorphisms? And is structure merely interchangeable with terms such as organization and system? Much of what is designated as structuralism in the behavioral sciences seems unquestioning "magic" structuralism (Boudon, 1971).

Boudon argues that the concept of structure has different meanings in different disciplines. While it stands for a scientific approach applicable to all the human sciences, its interpretations vary according to the context of intentional definition and the context of operative definition. In the former context, structure is purely terminological in that it is used to indicate distinctions which can usually be defined by other terms (e.g. system and organization). At most this indicates that one is dealing with a group of interdependent entities or a system of relationships. As an intentional definition, structure represents the stable and relatively unchanging characteristics of a system which control and limit change.

The operative context attempts to ascertain the structure of a particular phenomenon by means of a theory made up of axioms and theorems. This only becomes a structural analysis after the object has been demonstrated to be a system. That is, the two definitions represent sequential phases in which the subject matter is first analyzed for its system properties. As Boudon points out, the use of intentional definitions implies a theory of the interdependence of a system's elements. But often this does not produce an operative implementation either because the object has characteristics which thwart this step or the necessary tools of thought are not yet available. For Boudon, the difficulties and ambiguities of the concept of structure lie in the middle ground between these two phases of investigation.

He considers the structure concept to be a collection of homonyms rather than synonyms. In his view it is not possible at present to work with structural isomorphisms between different levels of organization. A structure is nothing more and nothing less than the theory of a system; and theories can be located at various levels of organization such that each theory is dependent on the characteristics of the system considered. Using only intentional definitions creates the mirage of structural isomorphism between levels of organization.

Are structure and organization synonymous? Not necessarily so, according to Boudon. Objects can be structured without at the same time being organized. In addition, the "organizations" through which behavior is expressed can be modified without corresponding changes in the "structure". Thus, organization does not always represent a synonym for structure. A distinction should be made between structure and organization so that the former refers to an apparent system of relations and the latter to a real system.

As a final step, we arrive at the top-most rung of Bunge's hierarchy: dialectical determination. Before doing so, it is necessary to distinguish interpretations of interaction from this category of determination.

Interaction and dialectical determination

Over the last decade developmental psychology has become increasingly concerned with the distinction between interaction and dialectical determination. There has been a general recognition that more adequate explanations of developmental processes are needed which break with simplistic notions of mechanical determination.

Interaction seems to take two general interpretations. In one, the term is used in a general sense to refer to how two or more independent variables combine or connect in their relationship to a dependent variable, a relationship that might be causal or noncausal. It is similar to statistical interaction in that it postulates nondirectional interaction. According to Olweus (1977), however, it is quite distinct, being typified more by the S–R approaches to child development (e.g. Bijou and Baer, 1961). In a second interpretation the notion of interdependency between elements is introduced as suggested by Bunge (1959). Elements can be described, located and measured independently of each other, and in this sense they are considered primary. However, as Previn and Lewis (1979) argue for their category of interdependent interaction, effects can only be understood in relation to each other which may be linear or non-linear. Having identified the elements of interest, the problem then is to account for the derived relations between them. In contrast to statistical interaction and interaction in the S–R sense, this form of interaction focuses on the question "How do organismic and environmental variables interact with each other?"

In an attempt to broaden the concept of interaction as it relates to the question "how?", developmental psychologists have recently turned to dialectics, a situation that was foreshadowed in Soviet psychology more than fifty years ago. Stemming from the idea of Heraclitus of constant change, through Plato, Aristotle and later Hegel and Marx, the meaning of dialectics has undergone successive changes so at the present time: "...the concept of dialectical orientation is often used in such an expansive and nebulous manner that the intrinsic power and attractiveness of dialectics may not find the fruition it deserves" (Baltes and Cornelius, 1977, p. 121). Despite this opinion there is general agreement as to the basic tenets and laws (see Riegel (1976) and Wozniak (1975)).

An important dialectical tenet is that activity and change are primary while the elements of a system are secondary or derived. According to this tenet one cannot separate organism and environment artificially from each other in order to isolate specific causes in development. The main task is to account for the nature of the progressive interactions between them. Another tenet emphasizes that development cannot be understood just in terms of efficient causes producing predictable outcomes; it is a probabilistic phenomenon which means that developmental progress can never be adequately predicted on the basis of individual elements. A further tenet contends that developmental progress can only be explained by studying it simultaneously along several separate but interrelated dimensions ranging from the biological level of organization through the social and cultural-historical levels to the outside physical world. In this respect, dialectics rejects any form of reductionism. Finally, there is the tenet that stability is momentary and change is continuous. Whenever developmental equilibrium is attained, it is quickly overtaken by conflicts and contradictions leading to further change. It is this tenet more than any other that marks it off from interdependent interaction.

Such dialectical tenets of development are covered by three basic laws: the law of the unity and struggle of opposites, the law of the transformation of quantity into quality, and the law of the negation of the negation. The third law holds that in development something old is replaced by something new (negation) which in turn is replaced by still newer properties (negation of the negation) but at the same time reinstating what is positive in the old at a higher level of functioning. Put another way, objects and phenomena have within themselves the conditions for their own destruction and transformation into a higher form. Any negation of the negation is treated as a moment in a ceaseless process of development. It is this law which biologists in particular find difficult to accept (e.g. Young, 1978), and it is one that is not given great prominence by contemporary dialecticians.

The second law asserts that when the intensity of magnitude of the phenomenon's intrinsic properties go beyond certain limits then quantitative change produces a qualitative or structural reorganization at a higher level. A new phenomenon therefore emerges controlled by a new set of laws. This law suggests that change may be discontinuous or saltatory rather than continuous or gradual in nature. Waddington's (1957) epigenetic landscape contains features which correspond with this law. Here the valley slopes are analogous to response thresholds or critical levels beyond which there will be an emergence of a new phenotypic function. In Waddington's terminology this means that the ball (i.e. a phenotype) is pushed over a watershed or catastrophe surface by internal or external forces into another developmental pathway. The impact of these agents is catalytic or consummatory with regard to an ongoing developmental process. Thus, the developing system involves discontinuities in which responding is not a continuous function of the controlling variables: when a threshold value is exceeded, gradual quantitative change directed by the controlling variables gives way to an abrupt discontinuity in functional quality.

According to the first law, stages or plateaux of equilibrium are not representative of development but rather continuous modifications occurring as a result of conflicts or contradictions between two or more dimensions. Equilibrium is a transitory event because the different dimensions are never at rest nor in perfect harmony with each other and yet in this continuing disharmony the system maintains its integrity as a whole. In other words, hierarchical integration persists despite ongoing differentiation. This law has been represented by the well-known triad thesis → antithesis → synthesis – but nowadays it is hardly ever referred to in dialectics perhaps because it depicts the role of contradiction too mechanistically. Modern thinking tends to side with Hegel's notion of radical self-contradiction which postulates not any kind of opposition passively received but rather an active, selective confrontation of opposites. This approach seems compatible with Whitehead's (1929) thinking on organic unity in which "every event reacts with every other, but not with all aspects of every other" (Waddington, 1978, p. 145).

The notion of contradiction, conflict or crisis, is central to dialectics, and represents the main source of development. In this respect, conflict between dimensions has no negative connotation, but rather contributes positively to further developmental advances. This feature of dialectics has assumed an important position in a number of approaches to development as it seems to offer a general causal explanation for developmental change.

The most comprehensive presentation of a dialectical perspective for developmental psychology has been given by Riegel (1976). In his view a dialectical theory of human development considers the simultaneous movements along at least four arbitrary but interdependent dimensions: inner–biological, individual–psychological, cultural–historical and outer–physical. This is a representation of the familiar levels of organization extended outwards in time, and further serves to mark off dialectics from interdependent interaction. Development is determined by the synchronization of any two dimensions and indirectly by all of them. But progressions along one dimension and between two different ones are not always synchronized, and the desynchronization results in confrontations which provide opportunities for further jumps in development. In Riegel's opinion most constructive developmental crises are brought about by asynchronies between the individual–psychological and cultural–sociological sequences. Developmental disruptions emanating from the inner–biological and outer–physical dimensions are considered to be relatively rare. The similarity with Vygotsky (1962) is obvious, but Riegel here seems to underplay the status of his inner-biological dimension. Pertinent examples can be found (e.g. the adolescent growth spurt) in which inner-biological changes can result in asynchronies with an individual-psychological dimension and consequently may lead to developmental change.

Concrete examples of dialectical determination applied to research on early development are becoming more frequent in the literature with the reappearance of transactional forms of analysis.

Transaction and dialogue: applied dialectics

Connected with the increasing popularity of dialectics in developmental psychology has been the reemergence of transactionalism as originally put forward by Dewey and Bentley (1949). According to Meacham (1977), the transactional model is a similar metaphor to dialectics in that relational activity and temporal change are primary, while the elements are secondary. However, dialectics represents a much more general world view to which transactional analysis contributes by examining real-life synchronizations suggested by dialectics. In this sense, the transactional model is also consistent with the principles of contextualism (Meacham, 1977).

The current enthusiasm for transactional analysis among developmental psychologists has been due, in no small part, to Sameroff and Chandler's (1975) reexamination of the purported consequences of early trauma. These authors contend that children with early traumas (e.g. delivery complications) do not necessarily become retarded unless such traumas are linked with poor environmental circumstances. It is the progressive transactions between organismic characteristics and environmental factors that lead to particular developmental outcomes, not single causes defined in terms of "nature" or "nurture". For example, the results of the multi-ethnic Hawaiian longitudinal study for all 670 children born on the island of Kauai in 1955 (Werner, Bierman and French, 1971) showed that perinatal complications were related to later physical and psychological deviations at 10 years of age only in conjunction with persistently inadequate environmental circumstances. This sort of finding led Sameroff and Chandler (1975) to propose a continuum of caretaking casualty as opposed to the more biologically-based one of reproductive casualty (Lilenfield, Pasamanick and Rogers, 1955). In accordance with Sameroff and Chandler's proposal, it is the complex interweaving of these two dimensions over time that has to be understood in order to explain developmental outcomes.

At an even more refined level of analysis are short-term temporal synchronies in communication processes. For example, in mother–infant interaction, Rosetti-Fereira (1978) distinguishes three interdependent components of dialogical transactions: syntony, synchrony and reciprocity. Syntony refers to heightened maternal sensitivity and readiness to perceive minor changes in her infant's behavior. The attainment of precise temporal synchrony between the mother's and her infant's behavior will depend on the level of maternal syntony. Further, the degree to which the mother–infant relationship is a reciprocal one in which the infant plays an active role will hinge on maternal syntony and her attempts to maintain an ongoing synchrony between her own time sequence and that of her infant. Together these three aspects of dialogical transactions form the optimal learning situation for the infant to perceive that his actions can have a controlling influence on his environment. Early distortions in the perception of control may lead to the expression of learned helplessness in social and non-social situations (Watson, 1977). Thus, the dialogues or synchronization of two time sequences, particularly between mother and infant, can be interpreted as the main driving force behind more general dialectical processes involved in early developmental changes.

Dialectics appears to integrate other categories of determination. It has been depicted as a superordinate orientation to theory construction which incorporates methodological pluralism (Baltes and Cornelius, 1977; Berman, 1977). On the other hand dialectics (particularly of the Marxist variety) emphasises the role of social-cultural conditions in development and such an emphasis may lead to an overly restricted interpretation of development.

Other criticisms of dialectics have been offered by Bunge (1982). In his view, dialectics has "a plausible kernel surrounded by a mystical fog" (p. 41). The fog is generated by the first (unity and struggle of opposites) and the third (negation of the negation) laws. As for the first law, the creative power of the struggle (or conflict) between opposites is open to question. Why should the qualitative change induced always have to go in the "right" direction (i.e. lead to a more advanced stage of development)? The same can be asked of the third law which also postulates progressive change. Dialectics offers no answer to this question. In fact it does not stress the possibilities that there may be stasis as well as regressive changes in development.

The second law (transformation of quantity into quality) is the plausible kernel according to Bunge (1982). Stated in the form that a stage is reached in every process when some new property emerges which in turn has its own mode of quantitative variation, it seems a reasonable (if vague) hypothesis. As such, it is perhaps the only dialectical law that is universally true. However, if this law is to have any real application then it needs to be formulated as a theorem in a general model of change.

Whether dialectics constitutes such a model or is just a metaphor for development is debatable. If, as some of its adherents claim, it is a model applicable throughout the life-cycle, then it is one of considerable generality. But this claim raises a problem in that the more general a model purports to be, the less it can be distinguished from a metaphor (Black, 1962). Thus, it is doubtful whether dialectics has the status of a theoretical model for explaining developmental change. Rather, it seems to be (particularly in the case of transactionalism) a metaphor that might be developed into a theory proper.

Conclusion

Aristotle, with his schema of four causes, attempted to find a minimum set of general explanatory principles that would make a given developmental process fully understandable. Over the succeeding centuries the explanation of organic change was reduced to Aristotle's category of efficient causation (particularly by Bacon), but by the twentieth century, the structure (if not the substance) of Aristotle's schema had been restored in the biological sciences. Now the assumption of pure causation (in terms of antecedent cause and consequent effect) was taken to be the result of a superficial analysis. At most, it applied only to a restricted range of subject matter: either to explain anomalies in the ordering of processes (Kuhn, 1977) or in their rate of change (Overton, 1984). Other types of explanation account for the orderliness and end-directedness of such processes.

Bunge's (1959) hierarchy of categories of determination has been used in this chapter to consider the other modes of explanation that have been commonly used in developmental theories. None of them alone provides a suitable basis for understanding developmental change since all share one common shortcoming. They fail to define the critical aspects of the organism-environment relationship that are important at different points in development. Unless this shortcoming is tackled, developmental theorizing will continue to be laden with explananda (e.g. maturation, equilibration, conflict) which are so general as to be almost empty of meaning.

To conclude, mechanistic causation is only one among several possible categories of determination which can be ordered according to increasing complexity. Modern science does not opt for one type of determination at the expense of others but recognizes that there is a spectrum of types of lawful determination that are needed. Biology not only uses teleology combined with statistical determination but also dialectical determination, structural determination, reciprocal causation and causal determination. There is no exclusive concern with final causes but rather an approach that spans the entire spectrum of determination. Developmental psychology, which ontologically presupposes biology, should also be characterized by the same approach.

Notes

1. The notion of non-materialistic teleology, "a sort of developmental soul" (Waddington, 1962, p. 99), was supported by Aristotle and the earlier epigeneticists despite their emphasis on empirical research. It was their only way of accounting for the apparent directiveness of development while at the same time denying preformationism. Their dogmatic vitalism served a useful purpose for it helped to establish the independence of biology from the excesses of Cartesian materialism and Newtonian science, and set the scene for a holistic approach to living organisms.
2. McDougall proposed three criteria for teleological or purposive explanation: some prospective reference to possible future events, reference to purposes or goals that are in some way ultimate or fundamental to an organism, and a number of subjective psychological categories (e.g. consciousness, foresight and desire). See Boden (1978), Chapter 2, for a lucid outline of McDougall's approach to teleological determination.
3. The flexibility of statistical determination *vis-à-vis* the rigidity of causal determination should also be contrasted with the plasticity of teleological determination. The contrast is one between a number of initial equipotential states leading to another condition (statistical determination) as against an initial state employing a number of means to reach a given goal (teleological determination).
4. A distinction should also be made between order and organization. Order is a qualitative statistical concept applicable to the law of entropy and organization is not. The latter is a physical attribute of a system while order refers to the regularity in arrangements or sequences of objects, numbers or events. Stating that a system is highly ordered conveys nothing about its level of organization. In entropy it is order rather than organization that tends to decrease in irreversible processes. Wicken (1979) provides an excellent discussion of the order-organization distinction.

References

Baltes, P. B. and Cornelius, S. W. (1977), "The status of dialectics in developmental psychology: theoretical orientation versus scientific method," in Datan, N. and Reese, H. W. (eds), *Life-span Developmental Psychology: Dialectical Perspectives on Experimental Research* , New York: Academic Press.

Bernstein, N. A. (1967), *The Coordination and Regulation of Movements*, London: Pergamon Press.

Berman, D. S. (1977), "Cognitive-behaviorism as a dialectic contradiction: the unity of opposites," *Human Development*, 21, 248–54.

Bijou, S. W. and Baer, D. M. (1961), *Child Development* (vol. 1), New York: Appleton-Century-Crofts

Black, M. (1962), *Models and Metaphors*, Ithaca: Cornell University Press.

Boden, M. A. (1978), *Purposive Explanation in Psychology*, Brighton: Harvester Press.

Boudon, R. (1971), *Uses of Structuralism*, London: Heinemann.

Bowlby, J. (1969), *Attachment*, New York: Basic Books.

Bunge, M. (1959), *Causality*, Cambridge, Mass.: Harvard University Press.

Bunge, M. (1982), *Scientific Materialism*, Dordrecht: Riedel.

Child, C. M. (1941), *Patterns and Problems of Development*, Chicago: University of Chicago Press.

Delbruck, M. (1971), "Aristotle-totle-totle," in Monod, J. and Borek, E. (eds), *Of Microbes and Life*, New York: Columbia University Press.

Dewey, J., and Bentley, A. F. (1949), *Knowing and the Known*, Boston: Beacon.

Gagné, R. M. (1968), "Contributions of learning to human development," *Psychological Review*, 75, 177–91.

Gibson, J. J. (1979), *The Ecological Approach to Visual Perception*, Boston: Houghton Mifflin.

Goodwin, B. C. (1985) "Constructional biology," in Butterworth, G. E., Ruthowska, J. and Scaife, M. (eds), *Evolution and Developmental Psychology*, Brighton: Harvester Press.

Guilford, J. P. (1967), *The Nature of Human Intelligence*, New York: McGraw-Hill.

Kagan, J., Kearsley, R. and Zelazo, P. (1978) *Infancy: its place in human development*, Cambridge, Mass.: Harvard University Press.

Klaus, M. H. and Kennell, J. H. (1976), *Maternal-infant Bonding*, St. Louis: Mosby.

Konner, M. (1972), "Aspects of the developmental ethology of a foraging people," in Blurton-Jones, N. (ed.), *Ethological Studies in Child Behavior*, Cambridge: Cambridge University Press.

Kronenfeld, D. and Decker, H. W. (1979), "Structuralism," *Annual Review of Anthropology*, 8, 503–41.

Kugler, P. N. (1986), "A morphological perspective on the origin and evolution of movement patterns," in Wade, M. G. and Whiting, H. T. A. (eds), *Motor Development in Children: Aspects of Coordination and Control*, Dordrecht: Martinus Nijhoff.

Kuhn, T. S. (1977), *The Essential Tension*, Chicago: University of Chicago Press.

Lilenfeld, A. M., Pasamanick, B. and Rogers, M. (1955), "Relationships between pregnancy experience and the development of certain neuropsychiatric disorders in childhood," *American Journal of Public Health*, 45, 637–43.

Longuet-Higgins (1969), "What biology is about," in Waddington, C. H. (ed.), *Towards a Theoretical Biology*, vol. 2, Edinburgh: Edinburgh University Press.

Lorenz, K. Z. (1965), *Evolution and Modification of Behavior*, Chicago: University of Chicago Press.

Løvtrup, S. (1974), *Epigenetics*, London: Wiley.

Mayr, E. (1974), "Behavior programs and evolutionary strategies," *American Scientist*, 62, 650–9.

Mayr, E. (1982), *The Growth of Biological Thought*, Cambridge, Mass.: Belknap Press of Harvard University Press.

McDougall, W. (1932), *The Energies of Men: a Study of the Fundamentals of Dynamic Psychology*, London: Methuen.

Meacham, J. A. (1977), "A transactional model of remembering," in Datan, N. and Reese, H. W. (eds), *Life-span Developmental Psychology; Dialectical Perspectives on Experimental Research*, New York: Academic Press.

Michel, G. F. (1981), "Right-handedness: a consequence of infant supine head-orientation preferences?" *Science*, 212, 685–7.

Nagel, E. (1961), *The Structure of Science*, London: Routledge and Kegan Paul.

Newman, S. A. and Leonard, C. M. (1983), "Against programs – limb generation without developmental information," in Fallon, J. F. and Caplan, A. I. (eds) *Limb Development and Regeneration* (Part A), New York: Liss.

Olweus, D. (1977), "A critical analysis of the 'modern' interactionist position," in Magnusson, D. and Endler, N. S. (eds), *Personality at the Crossroads: Current Issues in Interactional Psychology*, Hillsdale, N.J.: Erlbaum.

Oppenheim, R. W. (1981), "Ontogenetic adaptations and retrogressive processes in the development of the nervous system and behaviour," in

Connolly, K. J. and Prechtl, H. F. R. (eds), *Maturation and Development: Biological and Psychological Perspectives*, London: Heinemann.

Overton, W. F. (1975), "General systems, structures and development," in Riegel, K. F. and Rosenwald, G. C. (eds), *Structure and Transformation: Developmental and Historical Aspects*, New York: Wiley.

Overton, W. F. (1984), "World views and their influences on psychological theory and research: Kuhn-Lakatos-Laudan," *Advances in Child Development and Behavior*, 18, 191–226.

Piaget, J. (1971), *Biology and Knowledge*, Edinburgh: Edinburgh University Press.

Prechtl, H. F. R. (1977), *The Neurological Examination of the Full-term Newborn Infant*, London: Heinemann.

Prechtl, H. F. R. (1981), "The study of neural development as a perspective of clinical problems," in Connolly, K. J. and Prechtl, H. F. R. (eds), *Maturation and Development: Biological and Psychological Perspectives*, London: Heinemann.

Previn, L. A. and Lewis, M. (1979), "Overview of the internal-external issue," in Previn, L. A., and Lewis, M. (eds), *Perspectives in Interactional Psychology*, New York: Plenum Press.

Riegel, K. F. (1976), "The dialectics of human development," *American Psychologist*, 31, 689–700.

Rosetti-Ferreira, M. C. (1978), "Malnutrition and mother-infant synchrony: slow mental development," *International Journal of Behavioral Development*, 1978, 1, 207–19.

Salmon, W. C. (1975), "Theoretical explanation," in Korner, S. (ed.), *Explanation*, Oxford: Blackwell.

Sameroff, A. F. and Chandler, M. (1975), "Reproductive risk and the continuum of caretaking casualty," in Horowitz, F. D. (ed.), *Review of Child Development Research* (vol. 4), Chicago: University of Chicago Press.

Stent, G. S. (1981), "Strength and weakness of the genetic approach to the development of the nervous system," *Annual Review of Neurosciences*, 4, 163–94.

Stent, G. S. (1984), "Hermeneutics and the analysis of complex biological systems," in Degew, D. J. and Weber, B. H. (eds), *Evolution at the Crossroads: the New Biology and the New Philosophy of Science*, Cambridge, Mass.: MIT Press/Bradford Books.

Thom, R. (1975), *Structural Stability and Morphogenesis*, Reading: Mass.: Benjamin.

Vygotsky, L. S. (1962), *Thought and Language*, Cambridge, Mass.: MIT Press.

Waddington, C. H. (1957), *The Strategy of Genes*, London: Allen and Unwin.

Waddington, C. H. (1962), *New Patterns in Genetics and Development*, New York: Columbia University Press.

Waddington, C. H. (1978), "The process of evolution and notes on the evolution of mind," in Cobb, J. B. and Griffin, D. R. (eds), *Mind in Nature: Essays on the Interface of Science and Philosophy*, Washington, D. C.: University Press of America.

Watson, J. S. (1977), "Depression and the perception of control in early childhood," in Schulterbrandt, J. G. and Raskin, A. (eds), *Depression in Childhood: Diagnosis, Treatment and Conceptual Models*, New York: Raven Press.

Weismann, A. (1893), *The Germplasm: a Theory of Heredity*, New York: Scribner.

Werner, E. E., Bierman, J. and French, F. (1971), *The Children of Kauai: a longitudinal study from the prenatal period to age ten*, Honolulu: University of Hawaii Press.

Werner, H. (1948), *Comparative Psychology of Mental Development*, New York: International Universities Press.

Wener, H. (1957), "The concept of development from a comparative and organismic point of view," in Harris, D. B. (ed.), *The Concept of Development*, Minneapolis: University of Minnesota Press.

Whitehead, A. N. (1929), *Process and Reality*, London: Macmillan.

Wicken, J. S. (1979), "The generation of complexity in evolution: theoretical discussion," *Journal of Theoretical Biology*, 77, 349–65.

Wicken, J. S. (1981), "Causal explanations in classical and statistical thermodynamics," *Philosophy of Science*, 48, 65–77.

Wilden, A. (1972), *System and Structure: essays in communication and exchange*, London: Tavistock Publications.

Wozniak, R. H. (1975), "Dialectism and structuralism: the philosophical foundation of Soviet psychology and Piagetian developmental theory," in Riegel, K. F. and Rosenwald, G. C. (eds), *Structure and Transformation: Developmental and Historical Aspects*, New York: Wiley.

Young, J. Z. (1978), *Programs of the Brain*, Oxford: Oxford University Press.

2 Empirical evidence for causes in development

Peter Bryant

Two reasonably separable questions confront all developmental psychologists. One concerns the course of development. What, for example, are the differences between the intellectual skills of 3-year-olds and of 5-year-olds? Are there qualitative differences in the social responses of a child of four and the same child some years later and now a school child? The other is a causal question. Once one knows what these developmental changes are, one has to explain them and this means finding out what makes them happen.

Of course the two questions are not completely independent of one another, since evidence about one often suggests the form of the answer to the other. If you were to find for example that at a certain period in childhood there are striking changes in children's capacity to remember and then subsequent improvements in the way they tackle other tasks which could well depend on memory, it is a fair bet that changes in memory are responsible for these other intellectual developments. But it would be a bet only, not a certainty, and developmental psychologists should always avoid the trap of thinking that discoveries about the nature of a developmental change will also demonstrate what caused that change. They never do.

One has to work quite hard to bear this caveat in mind, because there is a marked imbalance in the amount of reasonable data that we have on the two questions. There is a great deal of rather impressive evidence about the first question. We argue about the details, but by now we really do know a great deal about the ways in which children's behavior changes as they grow older. It is easy to think of examples. There is considerable agreement about the nature of infants' perceptual and cognitive achievements, although new and exciting data are still coming in (Bremner, 1988). There is detailed knowledge about school children's failures and successes in several well tried logical and mathematical problems though the reasons for the pattern of these results is still a matter of intense argument (Fuson, 1988). The course of language acquisition is well charted (Ellis and Beattie, 1986). Our map of development is certainly incomplete and it is open to different interpretations, but there is a lot in it.

Evidence about causes is quite another matter. Though developmental psychologists have had a great deal to say about what causes the development that they are plotting, they have produced very little in the way of substantial evidence for their causal ideas. Piaget's (1975) well-known suggestions about what it is that prompts development are elaborate and take up a considerable proportion of his writing, and they have been influential particularly among educators. But his own empirical work barely touched on these claims. He did not test his causal theory, and though his colleagues did something towards filling that gap (Inhelder, Sinclair and Bovet, 1974) their work dealt with only a tiny proportion of the vast number of developmental changes which were charted in Piaget's voluminous research.

It is not just Piaget. Many quite strong causal claims are based on the wrong kind of evidence – on data which demonstrate what develops rather than what prompts the development. One example is Gelman and Gallistel's (1978) argument that the knowledge that children build when they learn to count lays the basis for much of their subsequent mathematical development. Their evidence, which is impressive in its own right, does nothing to establish the truth of this claim. They show that 3- and 4-year-old children have a better idea of what they are doing when they count than many psychologists had suspected up till then, and that evidence does indeed suggest their causal hypothesis. But it does no more than that. We cannot assume a causal connection between early counting and later mathematical development: we have to establish it.

I have picked on Gelman and Gallistel because their causal claims are so explicit, but once the point is made it is easy to see that we are all sinners – all of us, that is, who have ever done cross-sectional work. We assume when we look at perception in the neonate, at the way 9-month-olds search for objects, at the linguistic skills of the 3-year-old that what we observe is in some way connected to the child's later skills – that there is a link to be made, for example, beween the baby's search and his eventual fully-fledged knowledge about space. Because we think that the later skills in some way depend on the earlier ones, our ideas about these links are definitely, though often implicitly, causal. Yet have no right, as Sugarman has quite correctly pointed out (198, to make any such assumption. What a child does at a certain time (I am repeating Sugarman's argument) may simply be a kind of holding operation which will last until he learns a new pattern of responding. The only connection between the new and the old way of behaving may be that the one eventually replaces the other.

Why have developmental psychologists been so slow to test their ideas for such causal connections? What is the reason for the huge imbalance between the amount of convincing empirical work on the first question (a very great deal) and on the second (not all that much)? The answer is simple. It is far harder to gather convincing evidence about causes. Causal questions invariably face one with some very tough empirical problems indeed. In fact the problems are so great that we do not yet have an agreed and satisfactory solution to them. In my view the only reasonable way to find one is to consider first what makes causal question so hard for the research worker.

I have two stumbling blocks to suggest. One is the nature of most developmental theories, and the other the weaknesses of the main available empirical techniques when they are applied on their own.

The nature of developmental theories

Nearly all developmental psychologists have adopted the assumption, so effectively questioned by Sugarman, that one thing leads to another. Social behavior at one time leads in some way to a different form of social behavior at an older age, spatial understanding at an early age to a more sophisticated spatial understanding later on, and so on.

The consensus, however, ends at that point, for there are sharply different types of causal theories about these connections. In one type of theory the behavior in question produces its own development. The Gelman and Gallistel theory which I have already mentioned is a case in point. The child progresses from merely counting to knowing a great deal more about number and them hard to test. Let us take as an example the idea at the heart of Piaget's causal theory – the notion of equilibrium and disequilibrium. His idea of intellectual development as a series of discrete steps was based on the concept of equilibrium. A child remains at the same intellectual stage as long as his intellectual machinery gives him a coherent picture of what is going on around him. While this happens he is in "equilibrium". But inevitably, according to Piaget, the child is led into experiences which do not fit well into his scheme of things. The child finds that he has conflicting ideas about the same topic, and this conflict leads to disequilibrium, an unpleasant intellectual state which the child then does his best to resolve. The resolution which leads to a new equilibrium is the real cause of developmental change. To remove the conflict, to regain a coherent view, to reach a new equilibrium, the child has to acquire new and more sophisticated intellectual structures. The original disequilibrium is the cue to do all that.

It makes an exciting if stormy story, but how to test it? To do so one has to be able to say exactly what are the conflicting ideas that lead eventually to, say, the development of the ability to make logical inferences, and Piaget never does that. The very generality of the idea prevents its proof, and as I have remarked already there is very little empirical evidence from Piaget or his colleagues for conflict as a cause of development.

It is a rather similar story with Vygotsky's zone. Its causal implications have escaped the attention of research workers. Wertsch, McNamee, McLane and Budwig (1980) looked at the question of the zone by observing parents helping children solve a problem, and Brown and Ferrara (1985) by looking at transfer, but neither they nor anyone else have demonstrated that the help given by parents today leads to intellectual changes tomorrow. We have no direct evidence and once again the probable reason is the theories' generality. The help that one parent gives a 3-year-old child with a jigsaw is very different from the advice she offers to an older child solving a difficult arithmetical problem. Vygotsky does not tell us if there is an identifiable element common to both situations.

The plain fact is that theories of this type are hard, and in my view actually impossible, to test. In that case why were they adopted in the first place? Surely it would be more sensible to try a testable hypothesis.

It seems to me that the very nature of Piaget's ideas about developmental changes (the first question) forced him to adopt his *Deus ex machina* approach to the causal issue (the second question). Piaget always defined the early developmental stages negatively. The inevitable Genevan claim about anything that develops was always *"il y a trois stades"*, and in the first of these three stages the children were always completely at sea with whatever it is that they were being asked to do. Their behavior was described in negative terms, and that immediately removes a possible theoretical move. For you cannot say that the roots of a particular development are to be found in the child's earlier behavior if all that is to be observed in that early behavior is an unmitigated failure. Deficits cannot cause abilities any more than holes can cause hills.

So Piaget's, and perhaps Vygotsky's, preoccupation with what the younger child cannot do actually prevented them from looking to the child's early behavior for the experiences which lead to developmental changes, and they turned instead to external causes which remain beyond the bounds of empirical psychology.

Thus a negative account of the younger child's abilities forces you to turn to an indirect causal model. You cannot say that the child's early skills provide the basis for later development, because you are in effect denying that he has, at first, any such skills. You are forced into the position of arguing that his very lack of skill leads him into such a muddle that he has to be rescued or rescue himself with the help of some external factor – either a *Deus ex machina* or a deus in some other part of the machine. Since suggestions about these kinds of external factors are inevitably pitched in very general and untestable terms, I have to conclude that developmental theories which dwell on children's early incompetence are bound to yield causal hypotheses which are impossible to test in a detailed and satisfactory way.

What about theories which concentrate instead on more positive aspects of the behavior of young children? This sort of theory naturally leads to a quite different kind of causal model. If you discern an impressive and adaptive skill in younger children you are bound to wonder whether this skill provides the basis for some later development. The Gelman and Gallistel hypothesis is a clear example. These experimenters had concluded that the young child understands the basic principles behind counting and that naturally led them to argue that this basic understanding led quite naturally to the more sophisticated skill involved in arithmetic.

There is a familiar caveat to be sounded, however. Gelman and Gallistel did not, as I have already remarked, actually test their causal hypothesis. But there is at least a difference here, which is that they could have. The hypothesis is testable. In fact any causal hypothesis which explains later development as a product of some earlier measurable skill can be tested. We have of course to consider what is an adequate test – a topic that I shall discuss in the second part of this paper – but it is easy to see that it should in principle be quite easy to study the connection between two measurable aspects of behavior.

This view, given the nature of recent work on children's cognitive and social skills, is an optimistic one. For much of this work has stressed the competence of very young children. Work on infants is an obvious example, and it is interesting to see that research has already begun, with some impressive positive results, on the possible connection between the considerable perceptual skills of babies in their first six months and their intellectual development over the following years.

That is one example, and there are many other possibilities. The resistance to Piaget has produced many examples of apparently surprising skills in young children. I have already mentioned the considerable skills shown by very young babies (Bremner, 1988). It is quite clear as well that children lead an intricate social life (Kaye, 1982; Stern, 1977) from a very early age, and that their social skills spill over into their spatial (Butterworth, 1987), and into their cognitive, life (Wood, 1988; Perret-Clermont, 1980): one obvious example is the way that children share among themselves (Desforges and Desforges, 1980; Miller, 1984; Klein and Langer, 1987; Frydman and Bryant, 1988). The sophistication and speed of their early acquisition of language is widely recognized (Ellis and Beattie, 1986), and now we also know that there is more to linguistic development than just the business of learning to speak: the interest of 3- and 4-year-old children in word games and nursery rhymes (Maclean, Bryant and Bradley, 1987; Chukovsky, 1974), the poems that they produce (Dowker, 1988) and their sensitivity to rhyme (Lenel and Cantor, 1981; Bradley and Bryant 1983, 1985) demonstrate an active interest in language at a very early age.

This list could go on, but I will stop it at this point for fear of seeming to resort to an endless panegyric on children's intelligence, flexibility and resilience. That is not my intention. The point that I want to make is simply that we are in a very good position to test causal ideas that one thing leads to another, now that we know about the existence of a large number of sophisticated and measurable skills in young children.

Yet this opportunity has hardly ever been taken. There are, I think, two major exceptions. One is the concerted recent attempt to link work on infants' performance in perceptual tasks to the growth of their intellectual skills over the next five years or so (Bornstein and Sigman, 1986; Fagan and McGrath, 1981; Fagan and Singer, 1983; Rose, Slater and Perry, 1981). The other is the work, which I shall review later, on links between pre-school rhyming skills and reading. Otherwise the considerable body of recent work showing that children are much more proficient than we had thought tends to remain stubbornly cross sectional.

I think that there are two possible reasons for this hesitancy to make longitudinal links. One is that developmental psychologists may be afflicted by a kind of misguided learned helplessness: we may be under the illusion that the difficulties in testing causal ideas like Piaget's are difficulties for every causal theory. I hope that what I have written will remove that idea. Another possible reason is that it is still quite hard to get the methods for testing causal hypotheses exactly right. So I shall turn now to the methodological question.

Methods and causes

This will be a short section (short because I and my colleagues have rehearsed the argument in it several times elsewhere (Bryant and Bradley, 1985; Bryant and Goswami, 1987). My main message in it is to the person who is looking for one perfect method to test a causal hypothesis. My message is "Don't look for one perfect method". There is no such thing. The methods for pursuing causal hypotheses have strengths and weaknesses that are complementary, and thus a combination of different methods can make a powerful basis for a causal hypothesis.

The argument is simple. There are two main methods to

consider, the longitudinal method and the intervention study. Longitudinal predictions, provided that the measures are right and provided also that steps are taken to control extraneous factors like IQ and social class, can establish that a definite connection really does exist between an aspect of children's behavior early on and some development many years later. But such data do not demonstrate that the link is a causal one. The earlier and later skill may be related because both are determined by something else, something of which the experimenter knows nothing.

Intervention experiments, on the other hand, if they are done properly and include the right control groups, do establish causal connections. If you give children a particular experience which is not shared by the control group and if the only difference between the two groups is that one has that specific experience and the other does not, you can be sure that the experience is the cause of any difference that does emerge between the two groups. But the problem here is that you cannot be sure that this is what happens in real life. You have, in your laboratory, caused a change but it might be quite artificial and have nothing to do with the factors that prompt the development in real life.

These worries about the two methods would be depressing, were it not for the fact that each of them is strong where the other is weak. Longitudinal methods demonstrate a real connection, though they leave its causal nature in doubt. Intervention establishes that such a connection is causal, though it can never show that the connection really does exist. A combination of the two methods, therefore, is a powerful test of a causal hypothesis. Any study which is set up to test whether factor A determines a later development B, and shows both that A predicts B and that extra experience with A speeds up the development of B, can be taken as convincing evidence that A does affect B.

The trouble is that studies which combine the two methods in this way are rare. We ourselves set up such a study (Bradley and Bryant, 1983) to test our idea that the experiences which a child has with rhyme before she goes to school play a significant role when she begins to learn to read. We thought that rhyme introduces the child in a systematic manner to the way in which words can be divided into constituent sounds, and thus might help her when she came to learn to cope with alphabetic letters which represent those constituent sounds. In our rather large scale study we demonstrated that measures that we took before the children had begun to read of their sensitivity to rhyme were strongly related to their success in learning to read years later.

However, I think that it is fair to say that most other studies of causal factors have stuck to one method or to the other, and that is all one needs to say. It would serve no purpose to parade a series of longitudinal studies which on their own do nothing to eliminate the possibility of some unknown tertium quid, or to describe the many intervention studies whose results might have little to do with reality. We only need to note that the two approaches should be put together.

To summarize, I have tried to make four points. The first is that we must distinguish questions, and evidence, which are about what it is that develops, from questions about the causes of these developmental changes. Not to make this distinction is to run the risk of confusing evidence about one question with evidence about the other. Simply to show that you are right about the first question – what the developmental changes are – does nothing to show that your idea about the causes of the change are right or not.

My second point is that evidence on the causal side of developmental theories is still remarkably scarce, and that there is a very good reason for this marked imbalance, at least as far as traditional theories such as Piaget's are concerned. These theories dwell on young children's lack of skills, and therefore cannot make a direct causal connection between early and late behavior. That connection has to be made with the help of some external mechanism, but this mechanism inevitably has to be pitched in such general terms that it is quite untestable. Theories which concentrate on initial deficits in children's behavior offer a bleak prospect for anyone interested in establishing causal links empirically.

The third and more encouraging point is that most developmental research nowadays is not so concerned with deficits and lays great stress instead on the adaptive behavior of young children and the considerable skills which they have from an early age. Such evidence leads to a different type of causal hypothesis – that there is a direct link between the early behavior and subsequent developments. These hypotheses are much easier to test because they involve specific connections and are about measurable behavior. Yet so far rather little has been done on the causal side even with this new set of data.

My final point was about the best way to test such hypotheses, and I argued that the most convincing way to do so is with a combination of longitudinal prediction and intervention. We know how to test causal hypotheses and, at last, we have causal hypotheses that can be tested. I hope that developmental psychologists will no longer be so hesitant about this side of their work.

References

Bornstein, M. H. and Sigman, M. D. (1986), "Continuity in mental development from infancy," *Child Development*, 57, 251–74.

Bradley, L. and Bryant, P. E. (1983), "Categorising sounds and learning to read – a causal connection," *Nature*, 301, 419–21.

Bradley, L. and Bryant, P. E. (1985), Rhyme and Reason in Reading and Spelling. I.A.R.L.D. Monographs No. 1, Ann Arbor: University of Michigan Press.

Bremner, J. G. (1988), *Infancy*, Oxford: Blackwell.

Brown, A. and Ferrara, R. A. (1985), "Diagnosing zones of proximal development," in Wertsch, J. V. (ed.), *Culture, Communication and Cognition: Vygotskian Perspectives,* Cambridge: Cambridge University Press.

Bryant, P. E. and Bradley, L. (1985), *Children's Reading Problems*, Oxford: Blackwell.

Bryant, P. E. and Goswami, U. (1987), "Phonological awareness and learning to read," in Beech, J. and Colley, A. (eds), *Cognitive Approaches to Reading*, Chichester: Wiley.

Butterworth, G. (1987), "Some benefits of egocentrism," in Bruner, J. and Haste, H. (eds), *Making Sense*, London: Methuen.

Chukovsky, K. (1974), *From Two to Five*, Berkeley: University of California Press.

Desforges, A. and Desforges, G. (1980), "Number-based strategies of sharing in young children," *Educational Studies*, 6, 2, 97–109.

Dowker, A. (1988), "Rhyme and alliteration in poems elicited from young children," *Journal of Child Language* (in press).

Ellis, A. and Beattie, G. (1986), The Psychology of Language and Communication, London: Wiedenfeld and Nicholson.

Fagan, J. and McGrath, S. K. (1981), "Infant recognition memory and later intelligence," *Intelligence*, 5, 121–30.

Fagan, J. and Singer, L. T. (1983), "Infant recognition memory as a measure of intelligence," in Lipsitt, L. P. (ed.), *Advances in Infancy Research*, vol 2, Norwood NJ: Ablex.

Flavell, J. (1963), *The Developmental Psychology of Jean Piaget*, New York: Van Norstrand.

Frydman, O. and Bryant, P. E. (1988), "Sharing and the understanding of number equivalence by young children," *Cognitive Development* (in press).

Fuson, K. C. (1988), *Children's Counting and Concepts of Number*, New York: Springer Verlag.

Gelman, R. (1982), "Accessing one-to-one correspondence: still another paper about conservation," *British Journal of Psychology*, 73, 209–20.

Gelman, R. and Gallistel, C. R. (1978), *The Child's Understanding of Number*, Cambridge, Mass: Harvard University Press.

Inhelder, B., Sinclair, H. and Bovet, M. (1974), *Learning and the development of cognition*, London: Routledge and Kegan Paul.

Kaye, K. (1982), *The Mental and Social Life of Babies*, Brighton: Harvester Press.

Klein, A. and Langer, J. (1987), Elementary numerical construction by toddlers. Paper presented to the biennial meeting of the SRCD (1987) at Baltimore, USA.

Lenel, J. C. and Cantor, J. H. (1981), "Rhyme recognition and phonemic perception in young children," *Journal of Psycholinguistic Research*, 10, 57–68.

Lewis, M. and Brooks-Gunn, J. (1981), "Visual attention at 3 months as a predictor of cognition functioning at 2 years of age," *Intelligence*, 5, 131–40.

Maclean, M., Bryant, P. E. and Bradley, L. (1987), "Rhymes, nursery rhymes and reading in early childhood," *Merrill-Palmer Quarterly*, 33, 255–82.

Miller, K. (1984), "The child as the measurer of all things: measurement procedures and the development of quantitative concepts," in Sophian C. (ed.), *Origins and Cognitive Skills*, Hillsdale, N.J.: Erlbaum.

Perret-Clermont, A-N. (1980), *Social Interaction and Cognitive Development in Children*, London: Academic Press.

Piaget, J. (1975), *The Development of Thought: Equilibration of Cognitive Structures*, Oxford: Blackwell.

Rose, D. H., Slater, A. and Perry, H. (1981), "Prediction of childhood intelligence from habituation in early infancy," *Intelligence*, 10, 251–63.

Stern, D. (1977), *The First Relationship: Infant and Mother*, London: Fontana/Open Books.

Sugarman, S. (1987), "The priority of description in developmental psychology," *International Journal of Behavioral Development*, 10, 391–414.

Vygotsky, L. (1962), *Thought and Language*, Cambridge, Mass.: MIT Press.

Vygotsky, L. (1978), *Mind in Society*, Cambridge, Mass: Harvard University Press.

Wertsch, J. V. (1985), *Culture, Communication and Cognition: Vygotskian Perspectives*, Cambridge: Cambridge University Press.

Wertsch, J. V., McNamee, G. D., McLane, J. B. and Budwig, N. A. (1980), "The adult child dyad as a problem solving system," *Child Development*, 50, 1215–21.

Wood, D. (1988), *How Children Learn and Think*, Oxford: Blackwell.

Part II Causes of Development in Biology and Ethology

3   The causes of biological form

Brian Goodwin

This chapter is an enquiry into the causes of form in biology, which unites different aspects of the subject, from evolution to cognition, within a common perspective. Historical approaches to this problem have employed various dualisms to separate different parts of the processes that generate form, whether in organismic development, evolution, or cognition; and to locate different causes and characteristics in these parts, to one of which is usually ascribed a distinct essence which is regarded as the generator of distinctive form. A number of these dualisms are examined, starting with that of Descartes, out of which emerges an alternative approach to the problem.

The creator and the automaton

Descartes was of the opinion that such is the gulf between humans and animals that the behavior of the latter could be explained in purely mechanical terms, while humanity is possessed of a creative faculty, irreducible to mechanism, as revealed particularly in language. His definition of creativity was very perceptive. It involved essentially three components: unlimited variety, relevance or appropriateness, and freedom from stimulus control. A competent language user can generate a virtually unlimited variety of sentences, each of which is relevant or appropriate to the linguistic context, and the particular sentence selected for utterance is not dictated by an external controlling stimulus. Thus the criterion used by Descartes to distinguish between the human and the animal focused on creativity. Three centuries on, with a highly developed Cartesian science and a theory of evolution that was intended to account for the origin of species (hence of species differences such as speech), how has this criterion been solved, sharpened or transformed?

In the context of linguistics there is still much support for the view that speech is one of the most important properties by which Homo sapiens may be distinguished from other primates, despite the demonstration that the latter (chimpanzees, for example) are perfectly capable of learning a rudimentary sign language and using it creatively by combining signs in novel and contextually appropriate ways. Nevertheless Chomsky (1979), for one, insists that the extraordinarily rapid acquisition of linguistic competence by human infants and the degree of creativity displayed is so far beyond anything demonstrated by other species that it reveals a qualitatively distinct level of cognitive organization. He thus adopts a Cartesian stance on the issue, and the Cartesian criteria of creative expression are clearly elaborated and embraced in his *Cartesian Linguistics* (1966) and *Language and Mind* (1968).

On the other hand, Descartes' analytical principles for the study of automata, which for him included not only inanimate nature but all the phenomena of biology up to the level of the human mind, have resulted in a biological science dominated by mechanical explanation. Evolution, about which Descartes did not need to bother, is itself regarded as the outcome of a purely mechanical process of variation under natural selection which has generated not only nonhuman animals but also human beings, including their brains. So brains must also be mechanisms if one accepts the monistic mechanism of contemporary evolutionary theory, and whatever differences there are between humans and nonhumans must be of degree and not of kind. What has Chomsky to say about this? He takes the view (Piatelli-Palmarini 1980) that the human brain is an organ of thought and that, like other organs of the body, it differs from those of nonhuman animals because of the innate (genetic) differences between species.

Chomsky is fully aware that this innatism explains very little. It is a statement of a problem, not a solution, especially since the genetic differences between humans and chimpanzees amount to no more than about one per cent of their genomes (i.e. we are ninety-nine per cent the same, genetically). But in strictly conceptual terms, Chomsky is perfectly clear about the nature of evolutionary "explanations" that invoke natural selection to account for the development of differences of form and behavior between species, as he makes evident in the following: "It is perfectly safe to attribute this development to 'natural selection', so long as we realize that there is no substance to this assertion, that it amounts to nothing more than a belief that there is some naturalistic explanation for these phenomena" (Chomsky, 1968, p. 83). Since genes make molecules, genetics is a powerful tool for producing differences in the molecular composition of organisms, and for identifying the morphological, behavioral, or metabolic consequences of failing to make certain molecules. But it does not tell us how the molecules are organized into the dynamic, organized process that is the living organism.

Through the application of Descartes' principles for the quantitative reduction of complex systems to clear and simple elementary processes, it has emerged that animals are not the automata that Descartes believed them to be; and are in fact every bit as refractory to scientific understanding as the minds which Descartes singled out as the domain of irreducible creativity. The areas of biology that continue to defy a Cartesian reductionist analysis include brain function, embryonic development and the evolutionary origins of the major taxonomic groups of organisms. One could argue that these are precisely the areas of biology where creation is most in evidence. However, despite the clarity of Descartes' definition, creativity is perhaps not the best way of characterizing the nature of the problem with which we are presented in these aspects of organic nature. So let us see if we can come to terms with these properties of organisms and minds by a somewhat different approach. Transformation is at times the best way to seek resolution.

The problem of form

The three areas of difficulty identified above, namely brain function, embryonic development and the origins of the major classes of organism, have something in common: they all involve the generation of complex, organized forms. This is perfectly clear in the evolutionary origins of the major taxonomic groups (phylogeny), which are characterized by distinct morphological features; and in embryonic development (ontogeny) wherein organisms of specific form are generated from seeds, buds, or eggs. Behavior and cognition also involve the generation of ordered forms in space and time whether it be in play, ritualized courtship, pattern recognition or speech. These can all be regarded as the result of generative principles and rules of transformation operating together with the contingencies of context to produce appropriate forms. The problem is to identify the particular types of dynamic order that characterize evolving populations, developing organisms, and functioning brains, giving rise to the distinctive forms and patterns that constitute their natural expression.

This is the problem of form in biology. It is that part of the subject that has remained refractory to the analytical, reductionist tradition that Descartes did so much to promulgate and that has revealed so much about the molecular and cellular properties of organisms and brains. What it has not revealed is their dynamic organization at a level appropriate to the phenomena of form that are such a striking characteristic of the biological realm. In Kuhnian terms, this may simply be a puzzle, something that will eventually be resolved by the progressive accumulation of more detail; or it could be a problem, whose resolution requires a quite fundamental change of perspective and assumptions, amounting to a paradigm shift. Let me now briefly consider these alternatives, whose implications have been discussed in much more detail and from a variety of perspectives in two recent collections of essays (Ho and Saunders, 1984; Pollard, 1984). Although this may appear to take us on an excursion away from the focus of our enquiry into organisms and minds, it is necessary to clear the biological ground of certain conceptual obstacles. Once this is done, the consequences for an understanding of organized process in biology and the link with creative action should become clearer.

The biological dialogue

The dominant biological view of organisms is that they are complex self-reproducing systems whose specific properties have evolved by natural selection acting on spontaneous variation arising from gene mutation and genome rearrangement. In this description there are essentially two sets of forces acting on organisms: internal ones coming from the genome, causing variations in organismic properties (including form); and external ones coming from the environment, determining which of the variants survive and so are adapted. The organism itself is nowhere defined except as a self-reproducing entity, yet it is in some sense the broker that mediates between the internal, genetic forces and the external, environmental ones, acting so as to optimize the genetic stock. Generally, this mediation is taken to be direct in the sense that phenotypes are assumed to be determined or caused by genotypes, so that selection on the former leads to modification of the latter. Thus the organism is effectively a transparent shop window with genetic goods displayed directly to the naturally selective shopper, whose selection of appropriate articles ("characters") effectively creates the specific packages of goods we call the members of a species.

There are two fundamental dualisms in this description: between genotype and phenotype, and between organism and environment. I shall return to the second in the next section. According to the first, the former is considered to contain the essential causes of the latter. This is currently expressed by the metaphor of the program, applied to the set of genetic instructions which directs the construction of the organism during embryonic development by specifying which molecules are produced when, where, and in what quantities. The organism is thus held to be reducible to the molecules of which it is composed. And certainly the organism is, in biochemical terms, composed of nothing but molecules. The great achievement of molecular biology is to have elucidated the mechanisms whereby these molecules are made and their quantities controlled. The limitation of this description is that form is not, in general, explicable simply in terms of composition; nor in terms of composition plus a history of the particular conditions obtaining during the generation of the form out of its constituents. Water and ice have the same composition but quite different forms, which are not explicable by the statement that one form appears above zero centigrade and the other below. The explanation of form always requires a theory of organization, of how the constituents are ordered dynamically in space and time. This fact has been recognized from at least the time of Pythagoras, but it is frequently forgotten.

It is because of the absence of such a theory of the organism that both embryonic development and the evolutionary origins of the major taxonomic groups remain unsolved problems. No matter how much we learn about genes and molecules, ontogeny and phylogeny will not be understood until we have an exact description of the type of dynamic organization that characterizes the living state; just as the behavior of liquids could not be understood in a generative sense until there was a theory of the dynamic space–time order that characterizes the liquid state of matter.

One development in molecular genetics that emphasizes this point rather dramatically is the discovery that there is no correlation between the DNA content of species and their morphological or other complexity. Species of amphibia that are virtually identical morphologically nevertheless have great differences in the DNA content of their chromosomes, while as noted earlier, humans and chimpanzees, with very significant morphological and behavioral differences, are very similar in their DNA content. So it is not content or composition that counts, but organization. This point has been made over and over in the history of biology (see E. Russell, 1916) for a classical statement; and Goodwin (1985a,b) for recent analyses. But careers are not made out of wrestling with difficult problems. And the difficulties are most likely of our own making: we are looking at the problem the wrong way, identifying the wrong causes. The causal connections between genotype and phenotype are not simply atomic, Humean, cause and effect relations mediated by molecules. This duality, like the mind/body duality, generates confusion and mystification, and it has a similar origin (see Webster and Goodwin, 1982, for an analysis of the genetic program as an "Idea" or a formative "Soul", and the organism as the "Body").

Organism and environment

Let me now return to the second dualism on which is based the theory of adaptation under natural selection: between organism and environment. The scenario is that the environment preexists in the form of niches which pose problems for natural selection to solve by promoting organisms with appropriate characters for survival and reproduction in these niches. Spontaneous variations in the genotype result in phenotypic variations which constitute the raw material for this problem-solving exercise. From this perspective, natural selection tends to be seen as the formative or creative agent in the evolutionary process, providing organisms with specific forms and behaviors appropriate to currently prevailing environmental conditions of life. Once again, we see that the organism is a mediator of uncertain status between the genes, whose random variations cause random phenotypic variety (random in the sense of uncorrelated with environmental change), and the environment, whose pressures must be accommodated if the species is to survive.

The great insight of evolutionary theory is that organismic life-cycles undergo hereditary changes that depend upon a dynamic balance between influences internal and external to organisms, rates of change in populations being dependent upon these influences acting on constituent members of the population. The limitations arise once again from a failure to recognize the organism as an active agent with its own organizational principles, imposed between the genes and the environment. Organisms both select and alter their environments, and their intrinsic dynamic organization limits the hereditary changes that are possible, so that the variety available for evolution is restricted. There seems to be no other way of understanding the limited set of basic morphological types of organism that consitute the foundation of our systems of classification, nor of explaining why they nearly all appeared within the relatively very brief evolutionary period of the Cambrian, with very few fundamental innovations since (see, for example, Arthur, 1984; Reid, 1985). Furthermore, organisms themselves have the potential for appropriate response to the environment, so that much of the variation that is available for evolutionary change arises not from random genetic mutation but from the intrinsically regulative and plastic responses of the organism to the environment during its life-cycle. This plasticity can include genetic response, in the sense that environmental stress has been shown to result in adaptive changes in the genome in a number of plant species (Cullis, 1984). Thus the so-called creative power of natural selection is in fact very circumscribed (see Ho, 1985, for an analysis of these issues).

The extent to which competitive interactions are instrumental in shaping evolutionary changes is a further issue of current debate. Organisms are as cooperative as they are competitive (Bateson, 1986), and they make a living in a manner that usually poses no threat to ecological balance. The rather rapacious and territorial images of organismic life-strategies that dominate neo-Darwinist descriptions appear to be largely ideological projections onto the biological domain borne of a competitive and individualist society (Lewontin, Rose and Kamin, 1984). A more appropriate description for the evolutionary process than natural selection (which was of course derived from a comparison with the domestic selection of breeding stocks) is provided by the concept of dynamic stability. The environment does not select and shape organisms any more than a bath shapes the spiral form of the water as it flows down the plug-hole. Clearly, if there were no bath there would be no flow and no form; but what generates the details of the spiral pattern is a combination of the intrinsic properties of the liquid state of matter, together with all the contingencies operating on the dynamic process (height of water, size of hole, force of gravity, etc.). Neo-Darwinist descriptions tend greatly to exaggerate the role of the environment on the one hand, and the role of the genes on the other. Both of these undergo random (mutually uncorrelated) change. But organisms do not: they change in systematic and ordered ways, which is what makes taxonomy possible. Thus organisms, in a sense, turn randomness into order by virtue of their own principles of dynamic organization, as Waddington (1957) was fond of emphasizing. The evolutionary process is thus an exploration of the possibilities inherent in the living state, realized as organisms of specific form and function. "Adaptation" means no more and no less than the stability of a life-strategy, a dynamic process involving a set of transformations whose generic property is the repetition of a (life) cycle the period of which is the generation time. There is no organism/ environment duality in this process because the dynamic of the life-cycle extends across the boundary between the two. Organisms are, in thermodynamic terms, open systems. For example, there are developing marine organisms that generate electrical fields due to ion fluxes that extend beyond their structural boundaries, so that dynamically they are continuous with the environment, and similarly with other mass flows. We can, if we wish, separate different states of organization of matter, such as the living and the non-living, liquid and solid. But because one can transform into the other, the boundaries are always fuzzy, and the different states are united under transformation. Thus duality is replaced by state transition in a unified dynamic, so that there is no more of a duality between organism and environment than there is between bone and muscle in the organism, or between nucleus and cytoplasm in a cell.

The logic of process

The argument of this essay leads inexorably to the familiar proposition that life is process and transformation. The limitations of the dualities discussed above arise from the attempt to explain stability (of species or state of adaptation) in terms of something static and stable (genome or environmental niche) rather than as something dynamic (organism–environment cycle). The same applies to attempts to explain the stability of behavior (instinct, habit) or of cognitive activity (recognition, memory) in terms of stable "representations" or "internal models". All of these conceptual dualisms may be traced conceptually to the Cartesian philosophy of substance in which there are elementary things or objects (molecules, cells, organisms, species) which are acted upon by forces external to them, so that change arises from Humean atomistic cause-effect relationships between hierarchically ordered categories of objects constituted of more fundamental objects. This has the consequence that these things and the actions in which they are involved are all dead mechanisms because they have no life of their own. This was precisely Descartes' view: all such entities are in fact machines, automata. However, as we have seen, this view of organisms leads to numerous contradictions and difficulties because of the endless proliferation of dualisms that arise from any attempt to analyze process in static terms. Again, this insight is not new: Zeno instructed us in it many centuries ago. "There can be no doubt that the Humean conception of Causality ... must be wrong," write Harré and Madden in their book *Causal Powers* (1975). The alternative is to assert the primacy of process, so that change due to immanent power is of the essence whereas "things" maintaining stability of state are derived, and require explanation. We are thus led to dialectics, the logic of process.

A fully-developed theory of process has some quite startling consequences. If change is taken as primitive, then we must stop thinking about movement as something that happens to things as a consequence of forces from outside themselves acting within a preexisting space–time framework. Causality becomes immanent rather than transient, and what we call objects and their environments are self-generating complementary forms. There is no figure without a ground, and the only criterion of appropriateness is dynamic stability. Thus the meaning of a process is to be discovered simply by perception and experience of the complementary relationship between event and context. Space–time is an appropriate descriptive context for localized action connected with particular intentions, but it is generated and maintained by intention and action; it is not a preexistent given. The same is true of all types of stability: they are actively maintained and held by action which persists as long as the intention (holding in or on) persists, after which there is reversion to change. Thus everything transforms sooner or later and all is flux, but it is not chaotic. Process has its own logic. It is not classical two-valued logic, which runs into contradictions as soon as it is faced with processes that have properties of both continuity and transformation. What is required is a logic in which every value is an aspect of all values, by virtue of their primary inner connectedness, and in which there are no absolute and atomic, logical values as in the classical scheme (Jerman, 1986). Only thus is it possible to resolve the problem of primary relational order in space–time processes. Russell (1959) showed that classical logic, with the law of the excluded middle, is not compatible with a condition of such inner relation among the components of a dynamic system: for according to such logic, either the relation is a part of the nature of the components, or the relations are identical to the elements themselves. Neither alternative allows for a primary condition of inter-relatedness in which every "part" enfolds the whole (see also Bohm, 1980).

Fields and forms

However, relational order is precisely what characterizes the condition of organisms. As we have seen, it is not composition that determines organismic form and transformation, but dynamic organization. Classically, relational space–time order is described by fields, and field equations describe their dynamics. It is the absence of adequate field theories of organismic life-cycles and cognitive processes that accounts for the serious deficiencies in our understanding of organisms and minds, of evolution and cognition. Insofar as they currently exist, such theories of, say, embryonic development do give us some insight into the type of dynamic space–time order that could underlie the generation of biological form (Meinhardt, 1982; Oster and Odell, 1985; Goodwin and Trainor, 1985).

Furthermore, it appears that field descriptions come closest to embodying the logic of process described above. Harré and Madden (1975) have addressed precisely the question of how best to remedy the inadequacies of Cartesian or Humean causality, and conclude that an alternative can be derived from the field concept. They quote Faraday (1857) on the notion of force or power: "What I mean by the word [force] is the *source* or *sources* of all possible actions of the particles or materials of the universe: these being often called the *powers* of nature when spoken of in relation to the different manners in which their effects are shown." They then continue: "The 'lines of force' then picture the directional structure of powers or potentials, distributed in space. The fundamental entity then becomes a single, unified field, and in perpetual process of change as its structure modulates from one distribution of potentials of a certain value to another" (Harré and Madden, 1975, p. 175).

This vision of a single unified dynamic field, with different qualities and powers, goes well beyond what I have sought to describe in relation to the organic order. However, if we are to take seriously a dialectic of process, then this is where it leads us. And it is a far cry from the Cartesian world of mechanism. Whitehead (1929) put the distinction in the following condensed, if cryptic, form: "Descartes in his philosophy conceives the thinker as creating the occasional thought. The philosophy of organism inverts the order, and conceives the thought as a constituent operation in the creation of the occasional thinker ... In this inversion we have the final contrast between a philosophy of substance and a philosophy of organism."

If I understand it, the message here is that there are not things (e.g. thinkers) that generate thoughts; there are processes that generate complementary forms, such as thinkers and thoughts, together with all the other aspects appropriate to this dynamic constellation of phenomena. So mind is not in the brain, any more than life is in the organism. These are aspects of ordered processes that exist in the dynamic relationship of thinking and acting, cycling and transforming, generated across the moving, fuzzy boundary between inner and outer, subject and object. Life is relational order lived at the interface, where forms are generated. The developing embryo folds itself into layers that modulate the flux of its dynamic inner–outer order in characteristic ways in different tissues. The brain is a labyrinth of folded surfaces, a complex domain of mappings, projections, and transformations which create an unprecedented richness of relational experience between inner and outer, meaningful because of the complementarity of figure and ground, event and context.

What, then is the fate of Descartes' dualism in relation to organisms and minds? For him, the organism was a machine, an automaton. Our scientific culture has tried hard to validate this proposition. But the organism has resisted, just as the mind has resisted. And this resistance leads us towards a very different conclusion. An organism is a centre of immanent, self-generating or creative power, organized in terms of a relational order that results in a periodic pattern of transformation (a life-cycle) involving historical and actual components (genes and environment) and biological universals (the order of the living state). All living beings are both cause and effect of themselves, pure self-sustaining activity. They are a "natura naturans" rather than "natura naturata", creative rather than created, law-giving rather than lawful, makers rather than doers. But an organic philosophy of process forces us to the conclusion that, in a certain fundamental sense, much of this description applies as well to other aspects of the world as we know it (Watson, 1986). And so, in this sense, the world is also an organism, taking us both backwards to an earlier vision of reality as living process, and forward to a new appreciation of that vision. There are, of course, great differences between different aspects of this unified, living field, since there are local state transitions that result in the boundaries we use to distinguish different conditions of order. However, it is all unified under transformation. The current dialectic in biology leads to one of those startling changes of cultural viewpoint that brings self-generating power back into fundamental reality and banishes mechanism.

This conclusion is greatly strengthened by recent developments in physics, in which local causality has broken down as a basic explanatory principle and been replaced by a concept of non-local connectedness (Bohm, 1987). For instance, particles that were initially in a state described technically by a non-decomposable wave function and which then separate, such as two photons emitted from an atom, have correlated states irrespective of how far apart they are. Furthermore, a change in the measurement procedure applied to one photon, which affects its state, is immediately recorded in the state of the other, even if the measurement device is altered after the decomposition event so that the photons have "separated". Therefore these particles cannot be treated independently of one another, no matter how far apart they are.

These phenomena are direct consequences of quantum mechanics, recognized by Einstein, Podolsky and Rosen as long ago as 1932, but they have only recently been experimentally demonstrated. Bohm and Hiley (1984) have introduced a new field, the quantum potential, to describe these properties. But the conceptual understanding of such phenomena requires a shift in our basic view of reality which Bohm (1980) addresses with characteristic insight. The breakdown of mechanical causality as an explanatory principle in science presents us with the challenge of replacing automatic mechanisms with another conception of process and order. The conclusion of this essay is that the properties of organisms and minds as dynamic forms lead us in the same direction as that of basic physics: towards a conception of reality as a constellation of fields with immanent causal powers that generate characteristic states of ordered process. In biology, this perspective defines a research programme with a quite different emphasis to that deriving from the monistic mechanism of contemporary evolutionary theory, as has been described elsewhere (Webster and Goodwin, 1982; Goodwin, 1984, 1985). The challenge here is to find a solution to the problem of form in dynamic, transformational terms that unites history with order, creativity with intelligibility.

References

Arthur, W. (1984), *Mechanisms of Morphological Evolution*, Chichester: Wiley.

Bateson, P. (1985), "Sociobiology and human politics," in Rose, S. and Appignanesi, L. (eds), *Science and Beyond*, Oxford: Blackwell.

Bohm, D. (1980), *Wholeness and the Implicate Order*, London: Routledge and Kegan Paul.

Bohm, D. (1987), "Hidden variables and the implicate order," in Hilary, B. J. and Peat, F. J. (eds), *Quantum Implications: Essays in Honour of Daria Bohm*, London: Routledge and Kegan Paul.

Bohm, D, and Hiley, B. J. (1984), "Found," *Phys.* 14, 255.

Chomsky, N. (1966), "Cartesian linguistics," in *The History of Rationalist Thought*, New York: Harper and Row.

Chomsky, N. (1968), *Language and Mind*, California: Berkeley University Press.

Chomsky, N. (1979), *Language and Responsibility*, Brighton: Harvester Press.

Cullis, C. A. (1984), "Environmentally induced DNA changes," in Pollard, J. (ed.), *Evolutionary Theory: Paths into the Future*, pp. 203–16, Chichester: Wiley.

Goodwin, B. C. (1984), "Changing from an evolutionary to a generative paradigm in biology," in Pollard, J. W. (ed.), *Evolutionary Theory: Paths into the Future*, pp. 99–120, Chichester: Wiley.

Goodwin, B. C. (1985a), "What are the causes of morphogenesis?" *Bio Essays*, 3, 32–5.

Goodwin, B. C. (1985b), "Developing organisms as self-organizing fields," in Antonelli, P. L. (ed.), *Mathematical Essays on Growth and the Emergence of Form*, pp. 185–200, Alberta: University of Alberta Press.

Goodwin, B. C. and Trainor, L. E. H. (1985), "Tip and whorl morphogenesis in *Acetabularia* by calcium-regulated strain fields," *J. Theoret. Biol.*, 117, 79–106.

Harré, R. and Madden, E. H. (1975), *Causal Powers. A theory of Natural Necessity*, Oxford: Blackwell.

Ho, M-W. (1985), "Genetic fitness and natural selection: myth or metaphor," in *Evolution of Social Behavior and Integrative Levels*, 3rd T. C. Schneirla Conference, N. Y.

Ho, M-W. and Saunders, P. T. (eds) (1984), *Beyond Neo-Darwinism: An Introduction to the New Evolutionary Paradigm*, London: Academic Press.

Jerman, I. (1986), Some problems and perspectives in a dynamic understanding of life and organisms. Unpublished manuscript.

Leowontin, R., Rose, S. P. R. and Kamin, L. (1984), *Not in Our Genes*, London: Plenum Press.

Meinhardt, H. (1982), *Models of Biological Pattern Formation*, London: Academic Press.

Oster, G. and Odell, G. M. (1984), "The mechanochemistry of cytogens," *Physica*, **D12** 333–50.

Piatelli-Palmarini, M. (1980), *Language and Learning: The Debate between Jean Piaget and Noam Chomsky*, London: Routledge and Kegan Paul.

Pollard, J. W. (1984), *Evolutionary Theory: Paths into the Future*, Chichester: Wiley.

Reid, R. G. B. (1985), *Evolutionary Theory: The Unfinished Synthesis*, London: Croom Helm.

Russell, B. (1959), *My Philosophical Development*, London: Allen & Unwin.

Russell, E. S. (1916), *Form and Function*, London: Murray.

Waddington, C. H. (1957), *The Strategy of the Genes*, London: Allen & Unwin.

Watson, A. (1986), *The Birth of Structure: A Twentieth-century Copernican revolution*, Ph.D. thesis, University of Sussex.

Webster, G. C. and Goodwin, B. C. (1982), "The origin of species: a structuralist approach," *J. Soc. Biol. Struct.* 5, 15–47.

Whitehead, A. N. (1929), *Process and Reality*, Cambridge: Cambridge University Press.

4   Causes of development in ethology

Peter Slater

Introduction

To a seasoned ethologist my title will seem a strange one. Ethology is concerned with questions about how behavior develops, the mechanisms underlying it (often referred to as the study of causation), the biological function that it serves and its evolutionary history. Each of these four topics requires rather different approaches and the questions posed demand different answers. The ethologist thus tends to keep them rigorously separate, while the expression "causes of development" seems to confound two of them, or indeed even three if one adopts the terminology that the study of biological function is the search for "ultimate causes", while the study of mechanisms involves examining proximate causes.

Semantics aside, what I intend to do in this chapter is to give some account of the light that ethology has shed on mechanisms of development. For a long time ethologists showed rather little interest in development, but this has changed considerably in the last few decades. A great deal of very interesting work has gone on in that time, some of which is certainly of significance to developmental psychologists (Slater, 1980). There seem to me to be three main ways in which these ethological studies may have broader relevance:

1. The systems ethologists study are, by and large, simpler than those which concern psychologists studying human develop-
    ment. Whether one regards the evolutionary links between the
    two as likely to have led to common mechanisms is not of
    overriding significance here, for similar problems may also be
    solved in similar ways by convergence. Natural selection may
    have generated a mechanism in a very simple animal which is

conceptually similar to the solution thought out by a design engineer. The animal may thus be a simple model in which to study mechanisms and generate ideas and theories, which may then be tested in the more complex human case.

2. The functional perspective that ethology offers also has its role to play. To biologists the word function has a particular, rather specialized meaning (see Clutton-Brock, 1981). It refers to the selective advantage or survival value of a trait shown by an animal: in other words, the reason why natural selection has led to a feature of behavior. An interest in such matters is common to most biologists, but not one that attracts many psychologists. The application of such ideas to humans is obviously controversial. Again, however, the ideas may generate useful simple models. The functional perspective posits that animals behave in such a way as to "maximize inclusive fitness", that is to get as many copies of their genes as possible into the next generation, either through their own reproductive efforts or by assisting relatives. Some aspects of human behavior also fit in with this biological idea (e.g. weaning conflict, or patterns of wealth inheritance from males to their sisters' children rather than their wife's in societies where paternity uncertainty is great). But whether this sort of explanation is satisfying or convincing is very much a matter of taste. What the functional approach may do is to provide an interesting conceptual framework. We start by asking the question: "What is this behavior trying to achieve?" In a strictly biological argument, the answer will be "enhanced inclusive fitness", but for many aspects of human behavior it is more reasonable to think of other functions, to assume that the behavior is adapted to different, more short-term goals. It is still functional, still adapted, but to different ends which are not necessarily of biological significance. If one can suggest what those ends might be, the ways in which the behavior is adapted to them can be examined, as can the extent to which its form is optimal, in exactly the way that biologists interested in function approach their subject, but without necessarily assuming that natural selection has written the rules.

3. As biologists, ethologists can also offer a different perspective on behavioral development based on the insights that developmental biologists have gained into other biological

systems. Again, while behavior is obviously different from anatomy or physiology, most notably in the role that experience plays in its ontogeny, it may prove possible to generalize at the level of concepts and principles. In his recent book, *The Problems of Biology* (1986), John Maynard Smith remarked: "If this book were concerned only with unsolved problems of biology, 90 per cent of it would be devoted to two topics: behavior and development." That these are currently two of the most intellectually challenging areas of biology should perhaps give us hope that solutions to some of the apparently intractable problems they present may not be too far away.

These, then, are three reasons why a developmental psychologist can benefit from a grasp of what is going on in developmental ethology. Sadly, preconceptions abound about why such an effort might not appear worthwhile. Let us first dispel these.

Genes and development

Ethology and instincts
At the time when ethologists paid little attention to development, their main concern was with the mechanisms underlying behavior. The early theories of Lorenz and Tinbergen referred to concepts such as "fixed action pattern" (literally translated from the German as "inherited coordination"), "instinct" (their equivalent of the psychologists' drive) and "innate releasing mechanism". These ideas suggested a fixity about the behavior in which they were interested, as is indeed often true of the courtship displays of fish and birds on which early ethologists tended to specialize. The terms also suggested that inheritance was a preeminent influence in the production of such behavior. Indeed, the two things were often linked, with the idea of inheritance leading to invariance, while development which involved learning was thought of as prone to be rather variable. This is obviously a thoroughly naive view and may even be the opposite of the case, as I shall discuss later. But it had the effect of distracting ethologists' attention from development: they looked upon these fixed and genetically

determined behavior patterns as if they leapt fully-formed from the animal the first time they were called upon so that there was nothing very interesting to study about their ontogeny. Development was a subject for embryologists, not ethologists.

As well as leading to neglect of development, these ideas led to accusations of genetic determinism. Words such as innate and instinctive suggest that behavior patterns are unmodifiable and "blueprinted in the genes". For this reason, such words fell from fashion about twenty-five years ago when ethological studies of development really got going. The epigenetic approach, stressing the interactive nature of development, with the continual dialogue between the organism and its environment, left no space for simple notions of dichotomies between learning and instinct or nature and nurture. An aspect of behavior might normally appear fixed, but this did not mean that it could not be modified, nor indeed that a whole variety of environmental influences, including learning, had no role in its ontogeny. For example, normal visual input is essential to both amphibia and mammals if the visual system is to be wired up to achieve binocular vision (Blakemore and Cooper, 1970; Keating, 1974). This is obviously crucial for the development of behavior such as prey capture.

Because of the possibility of misunderstandings, it was wise of ethologists at that stage to adopt different terminology, though some objected that the word innate was still perfectly useable within an epigenetic framework to mean "unlearnt" or "not copied from others" or "present at birth" or "appearing the first time the animal is confronted with the appropriate stimulus". This highlights another problem with the word: it has so many potential meanings that its use can lead to confusions. It is therefore unfortunate that it has been reappearing in some recent ethological publications, usually in the sense of "not copied from other individuals" (e.g. Gould and Marler, 1984). Readers should certainly not assume from this that ethologists have returned to the rather naive genetic determinism of several decades ago: ethology has moved on a lot since then (see Halliday and Slater, 1983).

Sociobiology and genetic determinism
Sociobiology can be viewed as a branch of ethology, but one with a narrower perspective, as sociobiologists are interested, virtually

exclusively, in questions about function and evolution. Matters of causation and of development such as concern us here are just what they are *not* interested in. But this point is one that has escaped many of the critics of sociobiology, who are often geneticists and developmental biologists, and so have a very different perspective. The narrowness of the interests of sociobiologists has led them to be much misunderstood by those whose interests lie elsewhere. Why is this so?

To consider the evolution of behavior, and its biological function, assumes that the feature being discussed has some genetic basis, for natural selection works by changing gene frequencies. To be more precise, one should say that it depends on *variation* in the feature being based on genetic differences at least in part. This is because natural selection can only act by choosing between variants and its results can only be passed on if the variations we observe have some genetic basis. Without this, the behavior could not change from generation to generation. In addition, it is of course possible that the behavior, and the genes affecting it, do not vary *now* but they did so at some stage in the past so that natural selection led to the present situation.

The important point from this argument is that adaptation and evolution only require a feature to be affected by some genetic variation. As behavior is a complex phenomenon, which is undoubtedly polygenic in its origins, it is hard to imagine any aspect of it that does not fulfil this rather weak criterion. For example, the speed with which an animal runs in a running wheel is likely to be affected by a host of factors, such as visual acuity and leg length, all of which have at least some genetic basis to their variation. The results of behavior geneticists confirm this conclusion: wherever they have attempted to enhance some aspect of behavior by artificial selection experiments, this has proved possible, indicating that at least some of the variation in the behavior must be genetically based.

As far as sociobiologists are concerned this conclusion means that they are home and dry. Their theories require variation in behavior to have some genetic basis; they certainly do not require genetic determinism. This said, they often phrase themselves as if to imply the latter. A common shorthand in their arguments goes something like: "Take a gene A for altruism . . ." This certainly has a determinist ring about it. But for the argument to hold all it need imply is the existence of a gene that *affects the frequency of altruism* not one that *determines* whether it occurs or not. Thus, the shorthand may be misleading but the validity of the argument is not affected. The genetic requirements of sociobiology are actually rather slight.

All this said, sociobiologists should be more guarded in how they phrase themselves for, if they are misunderstood, the fault is very often their own. In a recent television program, Richard Dawkins, who is an expert at marketing ideas to a general audience, referred to an aspect of behavior as due to "an evolutionarily prewired program written by natural selection". Evocative words to the layman, but not ones likely to attract a developmental biologist! Although such phrasing is bound to create more heat than light, I would still argue that the disagreement over sociobiology is essentially a trivial one stemming from the fact that developmental biologists and sociobiologists are interested in different things.

The epigenetic approach

From the above discussion it should be clear that neither those ethologists who are interested in development nor sociobiologists, who are not, have any requirement for their theories to view genes as determining behavior. But, leaving aside these theoretical issues, what of the contribution of ethology to the study of development? Just as our understanding of the principles underlying morphological development has gained greatly from studies of frogs and fruit flies, so simpler systems can help us to understand the way in which behavior develops. To illustrate this I would like to take a classic case, but one that continues to yield insights: that of bird song development. This is an important topic, for it demonstrates both the subtleties of social and other forms of learning and also the importance of biological context to their understanding.

The development of bird song – a case history
From an early stage, studies of bird song have had an important influence on how ethologists view development. In all the songbirds so far studied, learning has been found to play a crucial

role in song development (see Slater, 1983, 1989 for a fuller account with references to the earlier literature). This learning is also, nearly always, restricted to the sounds of the species to which the young individual belongs, so that young robins learn to sing robin songs, chaffinches chaffinch song and so on. Despite the well known examples of mimicry between species, such as that shown by starlings and mockingbirds, in the great majority of cases copying is entirely within the species. In most species only males sing. If the young bird has appropriate experience, he will produce a more or less accurate copy of the songs that he hears; if he is denied that experience, he will only produce a rudimentary song which bears little relation to the normal adult song of his species. In the chaffinch, for example, the normal song is split into a series of phrases within which the notes are complex in structure and near identical in form. On the other hand, the young bird which is reared in isolation from adult song develops a much less structured song: it is about the right length, in the correct frequency band and split into a series of notes. But these do not have the normal complex structure nor do they fall into distinct phrases. Hearing adult males of the species singing is thus essential if song is to develop normally.

Findings such as these have led to what is often referred to as the "auditory template model" of song development (Marler, 1976). This views the young bird as hatching with a "crude template", which gives it a rough idea of what its own species song should sound like. This acts as a filter which is sufficient to prevent it from copying any totally inappropriate songs that it hears. As a result of hearing its own species song during the memorization phase, the template becomes sharpened up to become an exact one, a more or less accurate representation of the songs that the young bird has heard. This may occur before singing starts or may be a continuing process so that birds in full song can still modify their output in line with songs that they hear. Singing usually beings a year after the young bird hatches when, as a young adult, its testosterone starts to circulate. At this stage it listens to its own output, matches it to the exact template it has formed of its species specific song and eventually produces a near perfect copy of this. A bird that is deprived of the opportunity to hear adults of its own species can only base its song on the crude template with which it hatched, so producing a rudimentary effort. More extreme still, a young bird which is deafened can neither hear others of its own species nor monitor its own output: in this case the song is even more affected, often being little more than a screech.

This description of song development, involving learning, but rather severe limitations on what is learnt, obviously involves rather subtle interaction between the animal and its environment. The exact form of this interaction varies from species to species and, in a number of cases, the auditory template model has had to be modified as a result of recent work. This is especially true of the studies on swamp and song sparrows by Peter Marler, on whose earlier work the original model was largely based.

Marler sees the constraints on song learning as very substantial, on the basis of two lines of evidence. First, some young birds deafened at an early age still show aspects of song structure typical of their species: in swamp sparrows a series of identical syllables, in song sparrows a number of phrases of different syllables (Marler and Sherman, 1983). There may thus be a tendency to produce the appropriate structure without experience. Second, in swamp sparrows all songs are made up of a limited range of note types, falling into clear categories and with certain, apparently rather similar, intermediates missing. This has led Marler and Pickert (1984) to suggest that the young bird hatches with a repertoire of notes and that the learning process consists of selecting those that will be incorporated into song and placing them in the appropriate sequence. Such active pruning of the repertoire has implications for the nature of the constraints on what behavior develops. A young bird might fail to memorize a variety of sounds because it could not hear them or was in some other way insensitive to them. It might memorize them but fail to reproduce them because its vocal apparatus was not up to the task. Neither of these explanations can account for the rather limited repertoire of sounds that an adult swamp sparrow ends up producing. By studying the sounds that the young bird makes in sub-song (the period just before its full adult song crystallizes), Marler and Peters (1982) have found these to be highly varied, with many more notes than become incorporated in the final version of song. As the song crystallizes so the number is reduced by an active culling process, perfectly normal notes being rejected. Thus, the limitation is not on learning or memory, nor is it on production, but appears to be a much more active reduction; so far it is not clear why particular notes are lost and others survive into the adult song.

These data suggest that the "auditory template" idea is not an adequate description of the way in which sounds are selected for production from amongst the many that are heard. The template notion also implies a physical limitation within the nervous system of the young animal. Recent results suggest that something more subtle may be going on here. That young birds only learn the song of their own species in the wild may be partly that they have much closer contact with conspecifics than with birds of other species, so are much more likely to hear this song than others. But that alone could not account for the very precise match between the songs of successive generations, for many species often sing in the same small patch of wood. What is more likely is that social factors play an important part, young birds tending to learn from individuals with which they have formed a bond. There is now evidence from several species that young birds have to see and interact with their "tutors" before they will learn song from them (e.g. Payne, 1981). This is not true of all species: in most of the classic experiments on song learning the young birds were trained with tape-recordings. However, these newer results call into question quite a few of those obtained earlier. The fact that a bird will not learn a particular recorded song may not be because of incapacity on its part but simply because the stimulus provided has been inadequate.

One recurrent theme in studies of behavioral development has been that of "sensitive phases", embodying the idea that "the characteristics of an individual can be more strongly influenced by a given event at one stage of development than at other stages" (Bateson, 1979). In keeping with other aspects of ethological terminology, the phrasing has changed over the years, the bold and strong expression "critical period" giving way to the weaker, but more accurate "sensitive phase". As far as song is concerned, production starts, in the great majority of species, early in adult life. The timing of the memorization of song is much more varied, and illustrates the notion of sensitive phases particularly well. Even closely related species may show quite strong differences. Amongst the finches, for example, some learn only as juveniles, well before they start to sing themselves, some as both juveniles and as young adults, some only as young adults, and some can add new phrases to their repertoire throughout their lives. These differences in timing probably relate to variations in the function of song from species to species. Where it is a mate attractant, early learning is essential so that the young male can obtain a mate at the start of the first breeding season; where it is a signal between rival males on their neighboring territories, copying between them in adult life may be more appropriate so that birds on neighboring territories share signals. Functional thinking can help us a great deal to understand why particular developmental strategies exist.

Given that sensitive phases in song learning do exist, what are the underlying mechanisms involved? Although no young birds have been shown to learn in the nest, in some cases sensitivity has been found to rise shortly thereafter. This rise may be partly a feature of maturation in the brain and auditory system. In cases where the evidence comes only from field studies of song copying a delay in the start of song learning may simply arise because models are not available till later: the song season may be over by the time the young bird hatches or its father, the most likely song model at this stage, may not sing much while feeding his chicks. The evidence for rising *sensitivity* is not great.

More is now known about the end of the sensitive phase for song learning in several species. In some cases birds that reach adulthood without having learnt song have proved impossible to train thereafter, and this suggested that sensitivity ended at a particular point regardless of experience. However, most such evidence comes from birds trained with tape-recordings: could it be that they would have remained sensitive if the ideal stimulus of a singing male with which they could interact had been available? Recent results on zebra finches suggest that this might be so (Eales, 1987). In this species young birds normally learn song between thirty-five and sixty-five days of age; their song becomes fixed at the end of this period. Birds raised by females and so not exposed to male song before they start to sing, improvise a rather poor song. However, if a male tutor becomes available later, these birds can modify this to produce normal song long after the song of other individuals has crystallized. The end of the sensitive phase does therefore depend on experience in this species.

Rules of development

What do results such as those described above on bird song allow one to conclude about processes of development? As so often with ethological work, the major conclusion of a biologist is that nature shows marvellous diversity! It is certainly a lesson of recent ethology that natural selection has a remarkable capacity to generate different mechanisms beautifully matched to an animal's functional requirements. Simple generalizations based on one or two species are not therefore very helpful, and most of the early ones attempted have not proved lasting. It is more useful to examine a variety of cases separately in the first place and then attempt to build up a picture of how each mechanism found relates to the functions of the system. At this level generalizations may well be forthcoming once the properties of several different systems have been examined. What developmental mechanisms of this sort have ethologists proposed? The following are a few of the ideas that have been current in the past few years.

Sensitive phases

The study of sensitive phases in various systems, including that discussed above, falls nicely into the category of concepts about which generalizations are becoming possible (Bateson, 1981). It is satisfying that some of the properties of the sensitive phases found in bird song are similar to those found in other well-worked systems, such as imprinting. It is know that one of the brain areas involved in song learning (HVc: Nottebohm, 1984) is close to that where imprinting takes place in chicks (IMHV: Horn, 1986), HVc being the caudal part of the hyperstriatum ventrale, IMVH the intermediate and medial part; perhaps this is a sign of common underlying mechanisms. The two developmental systems certainly have several points in common. For example, in both cases the learning process shows a sensitive phase, timing of which varies in relation to function. In imprinting, the two processes of filial and sexual imprinting take place at different times the latter being delayed, in chicks, to a stage when siblings have attained adult plumage (Vidal, 1980). The early timing of filial imprinting enables the young one to learn about, and thus stay close to, its mother as soon as it is mobile and so liable to get into danger. The later timing of sexual imprinting is thought to enable the young birds to learn the characteristics of adult relatives which they can use as a yardstick in choosing a mate that is related to themselves to an optimal extent (Bateson, 1979). In both cases also the termination of the sensitive phase depends on experience during it. Sensitivity is retained longer in animals deprived of adequate experience. Conversely, some stimuli may be much more effective than others, and these are often those found in nature. Thus, just as swamp sparrows will only learn a very restricted range of syllable types so imprinting on a live hen is much more effective at abolishing the ability to show further learning than is an equivalent amount of experience of an artificial object (Boakes and Panter, 1985).

One of the reasons why the mechanisms underlying imprinting and song learning are similar may be that the two serve rather similar functions. Not only are both in birds, but both involve learning the characteristics of other individuals. In looking at development more generally similarities in points of detail are perhaps less likely to emerge and generalization will be harder to come by. The principles that emerge are thus likely to be at a more conceptual level as are those discussed in the next sections.

What shapes developmental trajectories?

In a series of papers, Bateson (1976a,b, 1983) has discussed the types of determinants that might be expected to affect development. Whether one is discussing genetic or environmental factors, he points out that these can vary from ones which have very specific effects to ones with generalized influence through much of development (Bateson, 1976a). This is certainly true, but it is not altogether clear how much further this notion gets one because, although he proposes a classification of determinants based on such distinctions, there are obviously no hard and fast lines between the general and the specific. Nevertheless, for anyone who needs reminding of the fact that development is a complicated business, it is a point to bear in mind.

Bateson (1983) has also made two other points which are, I think, worth repeating. He reminds us, as has Oppenheim (1981), that development is not simply a preparation for adulthood, though this is certainly part of it. Natural selection acts on young animals as well as adults; the animal that does not behave appropriately at each stage of its development will be selected out regardless of whether what it was doing at the time would have made it a better adult. This point is most obvious in species undergoing metamorphosis: the caterpillar will become a butterfly, but its behavior is quite different and certainly hardly a simple preparation for the adult state. Thus one must always remember that the behavior of a young animal adapts it to its own immediate surroundings as well as helping it to cope with the problems that confront it later in life.

At a more detailed level Bateson, partly following Gottlieb (1976), points out that the effects of different events or experiences on behavioral development can be split into those that are initiating, facilitating, maintaining or predisposing. Ideas such as these, concerning the way in which external events may change the course of development have been taken rather further in a recent paper by Chalmers (1987). He points out that ethologists have paid a good deal of attention to the rather abrupt changes involved in some development, concentrating on sensitive phases for example, but much less to the gradual changes which are undoubtedly more common. He explores the rules that may govern these, defining a rule as "a condition or set of conditions that help to govern the course of development". The essence of such rules, he argues, is that they involve feedback, the system monitoring some aspect of its output and adjusting its trajectory according to how it is doing.

This idea is not a novel one: control systems analysis has been found useful in many aspects of biology, as in psychology, and is obviously particularly appropriate to developmental situations in which the system can correct for disturbances. A good example is the weight gain in infant rats studied by McCance (1962). This normally follows a sigmoid curve but, if the young animal is deprived of food for a short time so that it is knocked off course, it does not then show a lower parallel trajectory but recovers to reattain the normal growth curve. The animal must therefore be monitoring its own progress against a set point which changes with age in the way that Chalmers proposes.

In this particular case weight is obviously assessed at a particular instant and compared to a set point that depends on the animal's age. But, in other cases the set point may be independent of age. The assessment may also vary: the system could monitor the rate of behavior at a particular time or the cumulative amount that has been performed during some preceding period. Chalmers suggests the sorts of experimental results which may help us to distinguish between which of these modes of assessment is operating. For example, if the set point in the rat weight gain example above had been independent of age the prediction would be that deprivation would lead to a trajectory parallel to the original one rather than returning to it.

Chalmers refers to rules of this sort as "directing rules" as they direct the course of development of an aspect of behavior. He differentiates these from "stopping rules" which lead to the termination of performance of one behavior pattern or phase of development. He delineates two main categories here: rules which are performance independent or performance dependent. This is a distinction we have already made when discussing bird song learning and imprinting. In both these cases there is evidence that whether or not learning has taken place may affect the timing of the end of the sensitive phase. These, therefore, are subject to performance dependent stopping rules. On the other hand, cessation of social play in kittens occurs when the mother cat ejects the young animals from the den. The timing of this event does not depend on the amount of play that has taken place, so this is a performance independent stopping rule.

Chalmers discusses many examples, largely from the development of primates, within this framework. This seems to me a particularly useful contribution, getting away from the rather desperate and fruitless idea that everything interacts with everything else during development, which has rightly been criticized (e.g. by Bateson, 1981), to asking much more precise questions about the exact processes involved.

The advantages of learning

To an ethologist, perhaps one of the most striking differences in development between aspects of behavior concerns whether or not learning plays a large part in the process. For example, why do the young of so many bird species have to *learn* their songs from other individuals? This is perhaps also a question of importance to the wider issue of the origins of vocal learning, which interests those concerned with the evolution of language (Nottebohm, 1972). One suggestion that has been made is that learning enables small groups to share a phrase or phrases which differ from other groups and so label individuals as belonging to the group (Baker and Cunningham, 1985). In this way birds may be able to choose mates appropriate to themselves, either of the same group or of other groups depending on whether inbreeding or outbreeding is advantageous. These ideas are contentious, however (see, for example, the diverse views expressed in the commentaries which accompany Baker and Cunningham's article), and it is not yet clear whether or not song dialects have any role in selective mate choice. What other reasons might then exist for birds to learn their songs? The following are three possibilities:

1. Learning may enable song to be matched to some aspect of the environment, as with the call note matching between members of a mated pair found in some finches (Mundinger, 1979). Neighbors often learn song from each other and, when countersinging on their territorial boundaries, tend to match the song types that they sing (Krebs, Ashcroft and van Orsdol, 1981). Song learning is likely to enhance the accuracy of this matching. It may also improve the match of the song to the environment. If singer and listener are some distance from each other, the sounds that travel best through the environment concerned are likely to be picked up and copied preferentially. There is evidence in some species that song structure does indeed differ between environments in an adaptive manner (e.g. Gish and Morton, 1981).

2. Where song is highly complex, learning may be the most economical way of achieving that complexity. Although this point is clearly speculative, it does appear that selection acts against increased genome size (Williams, 1966) and there seems little doubt that the production of a large repertoire would be less parsimonious to program without learning than with it.

3. The transmission of song from one individual to another through learning can be extremely accurate, though mistakes will certainly occur more frequently than do gene mutations. Given the complexities of development from fertilized egg to fully formed animal, and the precise transcription that would undoubtedly be required if fine details of song were to be passed on without learning, it is not hard to imagine that

learning may lead to greater accuracy (Slater and Ince, 1982). This suggestion does, of course, run counter to the old ethological view that "innateness" and invariance go hand in hand, but it seems nevertheless more plausible.

At present it seems, therefore, that learning is likely to have been selected as a mechanism of behavioral development either where the accurate transmission of detailed information is required or where the environment is unpredictable so that adaptation to it is best achieved by the young individual learning about its features. Conversely, behavior development that does not require learning may have been selected for where it is essential that the young animal responds correctly without experience. The animal that does not react appropriately to a predator the first time it sees one is unlikely to have a second chance!

Conclusion

In this chapter I have explored some of the ways in which ethologists have approached development in the recent past. I hope it is apparent that the naive genetic determinism that once plagued ethological thinking is now something of the past and that ethologists, using a variety of animal models, are coming up with interesting ideas about development. Hopefully these will complement those of developmental psychologists to provide new insights into what are, basically, different aspects of the same problem.

Acknowledgments

I am grateful to Drs Neil Chalmers, Lucy Eales and Andrew Whiten for helpful comments on the manuscript.

References

Baker, M. C. and Cunningham, M. A. (1985), "The biology of bird song dialects," *Behav. Brain. Sci.*, **8**, 85–133.

Bateson, P. P. G. (1976a), "Specificity and the origins of behavior," *Advances in the Study of Behavior,* 6, 1–20.

Bateson, P. P. G. (1976b), "Rules and reciprocity in behavioral development," in Bateson, P. P. G. and Hinde, R. A. (eds), *Growing Points in Ethology,* Cambridge: Cambridge University Press.

Bateson, P. P. G. (1979), "How do sensitive periods arise and what are they for?" *Anim. Behav.* 27, 470–86.

Bateson, P. P. G. (1981), "Control of sensitivity to the environment during development," in Immelmann, K., Barlow, G. W., Petrinovich, L. and Main, M. (eds), *Behavioral Development,* Cambridge: Cambridge University Press.

Bateson, P. P. G. (1983), "Genes, environment and the development of behavior," in Halliday, T. R. and Slater, P. J. B. (eds), *Animal Behavior, volume 3: Genes, Development and Learning,* Oxford: Blackwell Scientific Publications.

Blakemore, C. and Cooper, G. F. (1970), "Development of the brain depends on the visual environment," *Nature,* 228, 477–78.

Boakes, R. A. and Panter, D. (1985), "Secondary imprinting in the domestic chick blocked by previous exposure to a live hen," *Anim. Behav.,* 33, 353–65.

Chalmers, N. R. (1987), "Developmental pathways in behaviour," *Anim. Behav.,* 35, 659–74.

Clutton-Brock, T. H. (1981), "Function," in McFarland, D. J. (ed.), *The Oxford Companion to Animal Behaviour,* Oxford: Oxford University Press.

Eales, L. A. (1987), "Song learning in female raised zebra finches: another look at the sensitive phase," *Anim. Behav.,* 35, 1347–55.

Gish, S. L. and Morton, E. S. (1981), "Structural adaptations to local habitat acoustics in Carolina wren songs," *Z. Tierpsychol.,* 56, 74–84.

Gottlieb, G. (1976), "Conceptions of prenatal development: behavioral embryology," *Psychol. Rev.,* 83, 215–34.

Gould, J. L. and Marler, P. (1984), "Ethology and the natural history of learning," in Marler, P. and Terrace, H. S. (eds), *The Biology of Learning,* pp. 47–74, Berlin: Springer-Verlag.

Halliday, T. R. and Slater, P. J. B. (eds) (1983), *Animal Behaviour, volume 3: Genes, Development and Learning,* Oxford: Blackwell Scientific Publications.

Horn, G. (1986), *Memory, Imprinting, and the Brain: An Inquiry into Mechanisms,* Oxford: Clarendon Press.

Keating, M. J. (1974), "Visual function and binocular visual connexions," *Brit. Med. Bull.,* 30, 145–51.

Krebs, J. R., Ashcroft, R. and van Orsdol, K. (1981), "Song matching in the great tit *Parus major* L.," *Anim. Behav.,* 29, 918–23.

Marler, P. (1976), "Sensory templates in species-specific behavior," in

Fentress, J. C. (ed.), *Simpler Networks and Behavior*, Sunderland, Mass.: Sinauer.

Marler, P. and Peters, S. (1982), "Developmental overproduction and selective attrition: new processes in the epigenesis of birdsong," *Dev. Psychobiol.*, 15, 369–78.

Marler, P. and Pickert, R. (1984), "Species-universal microstructure in the learned song of the swamp sparrow (Melospiza georgiana)," *Anim. Behav.*, 32, 673–89.

Marler, P. and Sherman, V. (1983), "Song structure without auditory feedback: emendations of the auditory template hypothesis," *J. Neurosci.*, 3, 517–31.

Maynard Smith, J. (1986), *The Problems of Biology*, Oxford: Oxford University Press.

Mundinger, P. C. (1979), "Call learning in the Carduelinae: ethological and systematic considerations," *Syst. Zool.*, 28, 270–83.

Nottebohm, F. (1972), "The origins of vocal learning," *Am. Nat.*, 106, 116–40.

Nottebohm, F. (1984), "Birdsong as a model in which to study brain processes related to learning," *Condor*, 86, 227–36.

Oppenheim, R. W. (1981), "Ontogenetic adaptations and retrogressive processes in the development of the nervous system and behavior: a neuroembryological perspective," in Connolly, K. J. and Prechtl, H. F. R. (eds), *Maturation and Development: Biological and Psychological Perspectives*, Philadelphia: Lippincott.

Payne, R. B. (1981), "Song learning and social interaction in indigo buntings," *Anim. Behav.*, 29, 688–97.

Slater, P. J. B. (1980), "The relevance of ethology," in Sants, J. (ed.), *Developmental Psychology and Society*, London: Macmillan.

Slater, P. J. B. (1983), "Bird song learning: theme and variations," in Brush, A. H. and Clark, G. A. (eds), *Perspectives in Ornithology*, New York: Cambridge University Press.

Slater, P. J. B. (1989), "Bird song learning: causes and consequences," *Ethology, Ecology and Evolution*, 1.

Slater, P. J. B. and Ince, S. A. (1982), "Song development in chaffinches: what is learnt and when?" *Ibis*, 124, 21–6.

Vidal, J. M. (1980), "The relations between filial and sexual imprinting in the domestic fowl: effects of age and social experience," *Anim. Behav.*, 28, 880–91.

Williams, G. C. (1966), *Adaptation and Natural Selection*, Princeton: Princeton University Press.

Part III   Causes of cognitive development

5 The development of reasoning ability

P. N. Johnson-Laird

Introduction

Few people can doubt the importance of reasoning ability to human beings. Without it, there would be no mathematics, science or technology; no general principles, conventions, or laws, governing day-to-day social interactions; and, almost certainly, no culture and no art. The ability to reason is indeed one of the chief elements in human life that sets it apart from that of other animals, and that frees it from any narrow ecological "niche" in the environment. If you want to imagine how you would live if your species could not reason, look to the life of simple social mammals – and even they are not devoid of all powers of reason.

To reason is to derive new information from old: a systematic process of thought leads from one set of propositions to another. It may proceed from several premises to a single conclusion, or from a single premise to several conclusions. It may start from verbally expressed premises, or from a perceived or conceived state of affairs. It may yield a verbal conclusion, or a direct course of action. What sets reasoning apart from other forms of thought, such as imagining or creating, is that the process is based, usually tacitly, on principles that establish a particular sort of semantic relation between the premises and conclusion. Depending on the nature of these principles, the conclusion is a necessary, probable, or possible consequence of the premises.

Many of the inferences of daily life are rapid, automatic and unconscious. Without the ability to make these *implicit* inferences, written and spoken discourse would not function in their usual way. Consider the following brief text:

The pilot put the plane into a spin just before landing on the strip. He just got it out of it in time. Wasn't he lucky?

You have no difficulty in understanding what happened, yet in order to do so, you had to make several inferences. Every word in the first sentence is ambiguous, but you probably did not notice the ambiguities because they cancel each other out, e.g. "pilot" can refer to an expert navigator who guides ships into and out of port, "plane" can refer to a tree or to a tool, "spin" can refer to a drive, "landing" can refer to an area between two flights of stairs, "strip" can refer to the removal of clothes. You rapidly infer the appropriate meanings as you fit them together in order to establish a plausible sequence of events. Similarly, to make sense of the second sentence, you need to establish the correct antecedents for the pronouns, and you have to infer that the first "it" refers to the plane and that the second "it" refers to the spin. Of course, the inferences you make in construing the text are not valid. It is (just) possible that the story concerns a sailor who spun a woodworking tool before the boat he was guiding landed on the beach. The mechanism yields conclusions by default, that is, they are justified unless, and until, subsequent evidence controverts them. This provisional quality is the hallmark of implicit inferences.

The only secure form of inference is deduction, which is supposedly based on principles of logic alone. The purpose of these principles is to guarantee validity, and the fundamental principle of deduction is that a conclusion is valid only if there is no way in which it can be false given that the premises are true. Logicians have formulated calculi that capture the principles of valid deduction, and that enable inferences in many domains to be proved in a purely syntactic way, i.e. derived from the premises solely by the use of formal rules of inference (and in some formulations from additional axiomatic assumptions).

Since deduction is generally deemed to be the most important form of reasoning, it has been the focus of psychological theories. My aim in this chapter is accordingly to sketch an account of how deductive ability develops. There are always various theories in the literature, but my first task will be to argue that they are so seriously defective that it is not possible to believe in any of them. I take development to embrace both evolution and learning, and I shall try to draw a division of labor between them based on computational considerations that concern the tractability of different sorts of learning algorithm. My next concern will be the nature of reasoning ability itself. Part of the problem of the previous accounts of its acquisition, I shall argue, is that they are based on an erroneous view of what is supposed to develop. Theorists have assumed that children acquire some sort of mental logic containing formal rules of inference. A brief review of the evidence will show that this claim is at best equivocal, and that most forms of reasoning (apart from those explicitly derived from formal logic) depend on *semantic* processes. I shall outline such a theory, the so-called "mental model" theory, and then inquire into what has to develop if this theory is right. Finally, I will examine the implications of this approach for which aspects of reasoning are likely to have evolved, and which aspects are likely to depend on learning.

Three views about the development of reasoning ability

Psychologists have to confront the following puzzle: how is it possible for children to acquire the ability to make valid deductions when, until they have acquired this ability, they presumably have no access to any sort of logical processes? There are three main answers to this question to be found in the literature.

The first view is that the ability can be explained by the machinery of learning theory: certain responses, which are rational, will be reinforced, and other responses, which are not rational, will not be reinforced. The mechanisms of generalization and specialization will then fill out the repertoire of behaviors so as to produce a rational individual. I know of no serious attempt to explain in detail how this scheme would lead an individual to logical competence. At least one behaviorist, however, holds that the transition is possible: in B. F. Skinner's (1948) novel, *Walden II*, there is an intriguing moment when the hero translates a paper by his professor into logical symbolism and demonstrates that it contains a number of fallacies. Novels, are, of course, works of the imagination – or of whatever behaviorists have in place of it: an ability to make responses that have no previous history of reinforcement, perhaps – and so this particular fantasy should probably not be taken too seriously.

A more plausible attempt to exploit learning theory was made by Falmagne (1980). She suggested that children could acquire schemata corresponding to rules of inference by way of a concept-learning process. Children encounter instances of a given pattern of inference, receive feedback about it from other speakers or from reality, and then abstract the logical structure common to these experiences. Hence, given a modicum of reasoning ability, children could acquire further rules of inference by extracting them in this way from verbally expressed arguments. That this conjecture depends on an existing deductive ability is obvious: adults do not go about demonstrating their capacity to make valid deductions, and so it is necessary for children to be able to distinguish between valid and invalid inferences before they go to work with their "generalizing" and "abstracting" procedures. But, if children can determine for themselves that a given argument is valid, they already possess the very ability that the theory is intended to explain. It therefore appears best suited to accounting for the growth of logical ability under the influence of explicit feedback about the logical status of inferences. It cannot explain the origins of inferential competence.

The most influential answer to the riddle of logical development comes from the Swiss scholar, the late Jean Piaget. He and his collaborators aimed to understand the growth of knowledge from childhood, and to explain how children came to know the world, to be able to reason logically, and ultimately to participate in the development of formal logic and mathematics. The central assumption of Piagetian theory is that the mind develops according to the same principles that underlie the evolution of species – with the caveat that Piaget was never an orthodox neo-Darwinian. Like an amoeba, the mind is supposed to assimilate reality or else to transform itself to accommodate its environment. Indeed, evolution starts with the simple behavior of uni-cellular organisms, and eventually leads to the actions by which human infants come to cope with the world, and in turn to the mental representation of these actions, which, according to Piagetians, is the basis of all rational thought.

Piaget assumed that intellectual development is governed by one fundamental mechanism, an automatic tendency to self-regulation that seeks to establish an equilibrium within mental structures. This mechanism, which he called "equilibration", is triggered by internal conflict, and it operates outside consciousness or volition in order to reestablish a stable state. Each new equilibrium is the outcome of compensatory reversible operations, but each is also an occasion for further correction (see e.g. Inhelder and Piaget, 1964, pp. 292–3). Hence, development passes through a series of stages. During the first eighteen months or so of life, infants are primarily mastering simple sensory-motor skills. When they start to grasp the notions of cause and effect, they begin to develop language, and enter the second stage of intellectual growth, the "preoperational" stage. Language is a means for representing the world. Yet, long after the beginnings of language children still make fallacious inferences. If you ask them, for example, "Which are there more of, wooden beads or brown beads?", they answer "brown beads" even though all the beads are wooden. They fail to appreciate the necessary relation between the cardinalities of a set and its subsets. According to Piaget, the problem arises because the children have not understood the reversibility of certain operations. Only when they appreciate that the operation of forming the complement of a set is reversible will they realize that one set can be included in another. The grasp of this principle inaugurates the third stage of intellectual development, the stage of "concrete operations". The final mastery of deductive reasoning depends on the capacity to carry out such operations, not on concrete objects, but on the objects of thought, namely, abstract propositions. These procedures, which occur in the final Piagetian stage, the stage of "formal operations", are made possible by the development of a complete mental logic, which is supposed to correspond to the propositional calculus (see Inhelder and Piaget, 1958, p. 305).

Piaget was a great psychologist because he asked the right questions. However, his saga of intellectual stages is, alas, almost entirely a myth. The experimental evidence fails to substantiate it (cf. Bryant and Trabasso, 1971; Donaldson, 1978; Markman and Seibert, 1976; and many other findings that run counter to the hypothesized stages). More importantly, however, the theory has little explanatory power. I shall mention just four of its inadequacies:

1. If children have to internalize their actions, then what is it that controls those actions in the first place? If you answer "some sort of internal 'program' or plan", then the actions are already internalized.

2. To suppose that children fail to grasp that one class can be

included in another until they reach the stage of concrete operations implies that they cannot understand an assertion such as "robins are birds" until they reach this stage. But such an inability would make it impossible for children to learn the meanings of words until this stage of their development – and clearly children are able to master vocabulary with great facility. There is therefore an important distinction between the specific inclusions of Piagetian experiments and the general inclusion of semantics. Some of the "small print" in Piaget's works seems to allow for such discrepancies, but the distinction in content should be a major part of any theory of intellectual development.

3. Piaget never spelt out explicitly how the process of equilibration works. If he had done so, it would be possible to model the child's development from one stage to the next, and to simulate it in a computer program. That prospect remains remote. Moreover, the way in which children are supposed to acquire formal rules of inference remains entirely mysterious, especially in the light of my earlier remark that the acquisition must be carried out by processes that are unable to make valid inferences in a systematic way.

4. The Piagetian logic of formal operations is a considerable muddle. It is certainly not equivalent to the propositional calculus, and it is a considerable work of exegesis to determine what its logical status actually is (see Parsons, 1960; Ennis, 1975, 1978). As Braine and Rumaine (1983) remark in their magisterial review: "Piaget's logic clearly cannot develop at adolescence, or at any time: it is too problematic to stand as a psychological model of anything." And these commentators are defenders of the doctrine of mental logic!

Given these difficulties, it is not surprising to discover that the third view about reasoning ability is that it is innate. This thesis has been advanced by the philosopher, Jerry Fodor, in the context of his provocative claim that, apart from the acquisition of simple facts, all learning – whether of concepts, grammars, or logics – is impossible. Fodor's (1980) arguments for this most extreme version of nativism are ingenious, instructive, but hardly decisive.

Fodor starts with the hypothetical problem of how children who know propositional logic could learn the more powerful logic of the predicate calculus. The children have accordingly mastered the logic of such words as "and", "or", "not", and "if", which enable them to make such inferences as the following:

If it isn't raining, then John is playing in the meadow.
John isn't playing in the meadow.
And so: It's raining.

The predicate calculus calls for the use of quantifiers, i.e. such words as "some", "none", and "all", and it enables inferences of the following sort to be drawn:

If it hasn't rained at any time to-day, then some children will be playing in every meadow.
No children are playing in some of the meadows.
And so: It has rained at some time to-day.

Fodor argues that there is no way in which children can learn to make the transition from one calculus to the other. The reason, he says, is because to learn predicate logic the children will have to learn what is meant by such expressions as:

No children are playing in some of the meadows.

And this demand in turn requires the children to formulate within the conceptual apparatus available to them some such hypothesis as:

"No children are playing in some of the meadows" is true if and only if...

where the dots stand for a specification that captures the conditions in which the sentence would be true. But, says Fodor, such a hypothesis cannot be formulated within the conceptual apparatus available at the earlier stage, because that is precisely the respect in which the simpler logic is weaker than the powerful one. If children cannot even express the meanings of the powerful logic within the simpler one, then there is no way in which they can pass from one to the other.

The argument is quite general, and leads to the conclusion that complex concepts cannot be constructed out of simpler ones, because the simpler ones are not rich enough to allow the complex ones to be expressed. And thus ineluctably Fodor is drawn to his extreme brand of nativism.

Fodor's argument is valuable because it raises some important issues, but it is wrong. It must be wrong, since it proves too much, namely, that logic and concepts could not have evolved, either. The best way to establish this point is to consider development from a computational standpoint.

Evolution and learning from a computational standpoint

Imagine a computational device that is capable of self-reproduction, that has a basic armamentarium of procedures and methods for combining them, and that has causal connections with the world so that it is able to construct internal models of its environment on the basis of sensory information and to take actions guided by such models. It needs only a small number of inbuilt procedures, a working memory for the results of intermediate computations, and methods of combining existing procedures, in order to have the full computational power of recursive functions. It would thus be able to compute any function that can be computed. If this device were equipped with a learning algorithm, then it could put together complex skills by using its methods of combination to construct them out of its basic procedures. But what should the learning algorithm be?

From a computational standpoint, complex skills could develop solely as a result of evolution, since the capacity for self-reproduction enables a species to acquire new behaviors: a random shuffling of genes leads to modifications in the inborn patterns of behavior and natural selection filters out anything that does not confer an advantage to the species. This neo-Darwinian process can, in principle, be mimicked by a computational procedure for assembling new procedures out of old ones. However, any method of developing a new computational procedure by random variation is *intractable* in the technical sense that the number of possible combinations of existing procedures increases exponentially with the size of the desired program, and, beyond a certain size, there ceases to be any reasonable chance of discovering the required combination. This lesson was soon learned by those computer scientists who attempted to devise new programs using the neo-Darwinian algorithm (cf. Fogel, Owens and Walsh, 1966). The evolution of species is possible because there are millions of individuals engaged in reproduction over millions of years, and because a random variation does not have to assemble a viable morphology or behavior in one step *ab initio*: it merely modifies existing structures.

The neo-Darwinian algorithm is not the only one for learning. Many species have the ability to learn during their own lifetimes, which plainly enhances their adaptability. As I have shown elsewhere, the processes of induction can be reduced to a small set of primitive operations for generalization, but once again the size of the space of possible generalizations renders any general search algorithm intractable (see Johnson-Laird, in press). Hence, learning is tractable only if it is constrained either by features of the environment or by internal principles (or both). Algorithms based on internal constraints can use them to guide the initial generative stage and can therefore be tractable and efficient, though they may still depend on feedback from the environment. Such "neo-Lamarckian" algorithms depend, of course, on the prior evolution both of the constraints and of the overall design of an algorithm that permits their exploitation. Evolution must construct the necessary architecture for the initial neo-Lamarckian algorithms. Once established, the latter may themselves yield new algorithms of the same sort.

The boundary between the role of evolution and the role of learning can be drawn in terms of tractability. Simple abilities can be acquired using a neo-Darwinian algorithm provided that the environment constrains the process of acquisition. If the ability is not dependent on properties of the environment likely to vary, then the ability may have evolved, e.g. the algorithm for stereopsis, which depends on the constraint that one thing cannot be in two places at the same time (see Marr, 1982). If the ability depends on variable features of the environment, then it can be learned by trial-and-error given the innate endowment to exploit this neo-Darwinian algorithm, e.g. the acquisition of a simple adaptive habit. Abilities that call for highly complex procedures that need to be sensitive to the environment can be learned only if there are innate constraints on the process of learning. Once such constraints have evolved, it will be possible to mobilize them in an efficient neo-Lamarckian learning algorithm, e.g. the acquisition of language.

This analysis helps us to see why Fodor's argument goes wrong. If mental logic and concepts have evolved, then they have developed in a way that in principle can be simulated by a neo-Darwinian learning algorithm. At a certain point, a hypothetical computational device could have evolved the propositional calculus, and we can then ask how it could extend this capacity to the predicate calculus. If Fodor were right, then there should be no learning algorithm – not even the neo-Darwinian one – by which this transition could be made. Hence, the predicate calculus could not evolve.

The source of Fodor's error is his assumption that a system that has mastered only the propositional calculus cannot represent the truth conditions of statements from a more powerful logic. There are no grounds for making this assumption: logical power is not equivalent to representational power. A computational device with the power of recursive functions can represent the predicate calculus, but it may be at a stage where it can reason only propositionally. It does not follow that it is unable to learn the predicate calculus. It can in principle put together its basic functions so as to construct the truth conditions of statements within the calculus and rules of inference for it. In short, the system has a certain degree of *computational* power, which will not increase if it is already the power of recursive functions. Yet, it can learn to increase its *logical* power. The system may have the power to represent any logic or any concept, but it will lack the logical power of a particular calculus until it has constructed it by forming the right combination of basic building blocks. It is therefore possible to acquire a more powerful logic and to learn richer concepts.

The mental model theory of reasoning

Before a sensible theory of the development of reasoning can be proposed, it is necessary to ask oneself this question: what is it that develops as children acquire the ability to reason? The answer has nearly always been: a logical calculus. Whenever philosophers, logicians, or psychologists ponder on this mystery, they think about it in a conscious and deliberate way. This serial mode of thought, combined with literacy, appears to inculcate in every domain of knowledge a concern for symbolic manipulations according to formal rules. Indeed, it is this mode of thought that is responsible for the development of the study of logic. Once the subject exists and provides a calculus for establishing the validity of inferences, there is nothing easier than to postulate that some mental embodiment of the calculus guides the process of thought itself. Theorists as diverse as Boole, Mill, Henle, Chomsky, Davidson, Piaget, Fodor, Braine and Rips have all in their different ways, sometimes implicitly, assumed that assertions have a logical form and that thought depends on rules for manipulating logical forms – in other words, it depends on a mental logic.

In fact, the assumption is groundless. The principles that govern the inferences that you make in daily life are not accessible to introspection, unless you are one of those singular individuals who tries to put into practice what you have been taught in a logic course, and there is no empirical evidence that unequivocally supports the use of formal rules of inference. Such evidence that there is about reasoning runs quite contrary to a formal theory. It shows that the content of the premises can have a decisive effect on what conclusion an individual draws (see e.g. Wason and Johnson-Laird, 1972). Even Piaget made this discovery, which he characteristically explained away by his concept of the "horizontal décalage", which seems to be little more than a redescription of the phenomenon. Yet the phenomenon is inimical to formal theories of inference. The evidence also shows that when people reason they are concerned with matters of truth and falsity. They are biased towards verifying hypotheses (see e.g. Bruner, Goodnow and Austin, 1956; Wason, 1968). They are also biased by their beliefs about what is true, and these prejudices influence both the conclusions that they draw for themselves (Oakhill and Johnson-Laird, 1985a) and their evaluations of given conclusions (Evans, Barston and Pollard, 1983). Semantic informativeness likewise appears to govern what conclusions people draw spontaneously. They do not throw semantic information away even if the resulting conclusions are logically valid. Hence, given the premise:

John got out of bed

no sane individual concludes:

John got out of bed or Fermat's last theorem has been proved (or both)

though this conclusion follows validly from the premise.

The burden of these observations is that reasoning is a semantic process, not the quasi-syntactic manipulation of formal uninterpreted symbols. This principle lies at the heart of the "mental model" theory of reasoning (see Johnson-Laird, 1983). This theory assumes that deductive inferences are made on the basis of three main steps:

1. The reasoner imagines how the world would be if the premises were true, taking into account any general knowledge that is triggered by their interpretation. This representation is remote from the linguistic form of the premises and consists of a mental model of the relevant state of affairs, i.e. a model that is close in form to the perception of the relevant events, with a mental token for each relevant individual, property and relation.

2. The reasoner formulates a novel conclusion that is true in the model, i.e. a conclusion establishing something that was not explicitly presented in the premises. If there is no such conclusion, then the reasoner considers that there is nothing that follows from the premises.

3. The reasoner attempts to construct an alternative model of the premises that refutes the conclusion drawn in the previous step. If there is no such model, then the conclusion is valid. If there is such a model, then the reasoner returns to the previous step, and attempts to formulate an informative conclusion that is true in all the models so far constructed. If the reasoner is uncertain about the existence of such a model, then the conclusion is drawn on a tentative basis.

This theory, which is part of a larger theory of higher processes, has been corroborated experimentally for a variety of forms of deduction, including children's syllogistic reasoning (Johnson-Laird, Oakhill and Bull, 1986), and a number of computational models of it have been implemented. The theory also makes sense of two other phenomena: the implicit inferences that occur in comprehension, which I described earlier, and the vagaries in the logical force of conditional assertions (see Johnson-Laird, 1986). Both these phenomena count against formal notions of human reasoning. The empirical results have yet to be explained by any theory based on logic; and there are so far no empirical findings that count against the mental model theory of reasoning.

If the mental model theory is correct, then what has to develop as children acquire the ability to reason deductively? There are two essential skills. The first is linguistic ability. Children need to be able to understand discourse by constructing representations of the states of affairs that it describes (i.e. mental models), to formulate descriptions of the situations that they have encoded in mental models, and to evaluate the truth value of assertions in relation to mental models. These linguistic skills are all ultimately dependent on a tacit grasp of the concepts of truth and falsity.

The second component skill that children must acquire is the ability to search for counterexamples. It is this skill alone that is needed, over and above the ability to use language, if an individual is going to be able to make deductive inferences. Hence, one of the virtues of the mental model theory of reasoning is that it reduces the explanatory load for developmental theory. The reasoning algorithm that I outlined above calls for only one stage – the search for models that refute a conclusion – that is peculiar to the process of deductive reasoning. There is no need to explain how formal rules of inference are acquired by children, because the theory has no recourse to them in accounting for the ability to reason.

The semantic development of language

The evolution of language is a topic beyond the scope of this chapter, but I want to consider some of the main landmarks in its semantic development, where I shall at first put to one side whether a particular element is a consequence of evolution or learning from experience. My aim is to establish some principles that are relevant to reasoning ability.

In order for the mind (or a computational device) to be equipped with a language, it has to be able to make symbolic responses and to interpret them. One individual can then communicate to another, and members of the community can thus experience the world by proxy: they can imagine how the world is, not on the basis of direct perception, but from their ability to construct models based on symbolic behaviors. A precursor to communication is accordingly the capacity to construct mental models. Another precursor is the ability to keep track of the status of such models, since the mind must distinguish between those that derive from perception and those that derive from communication and that therefore represent possible states of affairs. One mechanism that may underlie this development is the mind's capacity to construct models of its own abilities. Its access to a model of its ability to construct perceptual representations could play a central part in enabling it to grasp the idea that models represent *possible* states of affairs.

Still another precursor to language is the ability to set up goals, i.e. models of possible states of affairs, and to act so as to achieve them. Access to a model of this ability again appears to be crucial to the development of intentional behavior. And this ability, in turn, underlies the use of language to refer: a speaker intends to refer to something, and is understood once the content of this intention has been recovered (see Grice, 1957).

A further component of referential language is the competence to compare models based on its interpretation with those based on perception or memory. If the model of an assertion forms part of a model based on other evidence, then the assertion is true as assessed by that evidence. Each of the individuals, properties of individuals, and relations amongst them, can be carried over intact from the first model to the second model, though the latter may embody much else besides (cf. Kamp, 1980). It is part of linguistic competence to be able to make such a comparison and to act on its results; but it is a higher-order metalinguistic skill to know what one is doing and to be able to refer to the outcome. This latter skill provides the organism with an explicit knowledge of truth and falsity, and it will thereby be able to use the predicates "true" and "false" in an appropriate way. The skill depends on access to a model of the lower-level competence. A model of this skill may, in turn, be embedded within a still higher level – a meta-meta level that is reflected in the capacity to construct theories of semantics.

One of the striking features of development is, as we have just

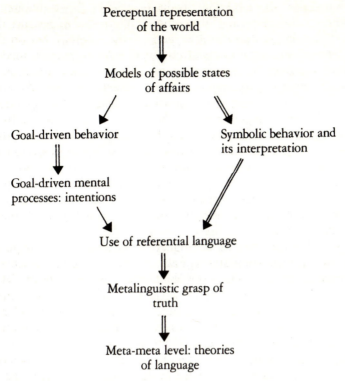

Figure 5.1   Landmarks in the semantic development of language

seen, the regular relation between a skill at one level and the subsequent formation of an explicit representation of it. The mechanism of forming an embedded model of the mind's competence at one stage within its competence at a later stage appears to be ubiquitous. This regularity suggests that mental architecture contains in-built machinery that allows the mind to form models of its own abilities, and even models of this ability itself. I have elsewhere characterized this mechanism as the recursive capacity to embed mental models within mental models (see Johnson-Laird, 1983, Chap. 6). It appears to underlie the development of language, reasoning, meta-cognitive skills, intentional behavior, and indeed our ability to introspect on our own aptitudes and proclivities – to examine, in short, models of ourselves. It may even be a sensible way to construe Piaget's notion of "internalization": the organism gains a high-level access to a model, necessarily incomplete, of some low-level capacity.

I have summarized the major landmarks in the semantic development of language in Figures 5.1. How does this system develop? In particular, what has its origin in evolution and emerges as a result of an individual's maturation? And what depends on the individual learning from experience?

The unfolding of the linguistic system, and particularly its meta-linguistic component, appears to be governed by an innate program. Children need experiences of their language in order to acquire it, but they do not have to construct the linguistic system *de novo*. Likewise, various components on which the linguistic system depends similarly appear to emerge as a result of maturation. In particular, working memory develops in capacity throughout childhood (Case, Kurland and Goldberg, 1982; Hitch and Halliday, 1983), but the system itself is not learned: it is a part of the fundamental architecture of the mind, since such a memory for the intermediate results of computations lies at the heart of computational power.

There is considerable biological evidence about the development of the linguistic system. For example, there are known anatomical structures (the vocal apparatus and areas of the brain) that are intimately concerned in the use of language. The pattern of linguistic development shows striking regularities from one culture to another. The effects of brain-damage, rearing in isolation, and certain maladies, all suggest that the linguistic system unfolds within a critical period, and that children can pick up their native tongue with little beyond mere exposure to it. There are therefore grounds for assuming that the architecture of the linguistic system has been acquired as a result of evolution. This architecture includes the landmarks shown in Figure 5.1. Four decisive evolutionary steps were necessary: first, the development of a system to represent the world on the basis of energy impinging from distal sources; second, the emergence of a working memory for the intermediate results of computations; third, the development of communicative behaviors that could convey the contents of mental models, and, fourth, the development of the ability to form models of the system's own capabilities. The first of these steps appears to have occurred in the evolution of organisms with visual systems that can construct three-dimensional representations of the world. The second step was a precursor to the ability to learn from experience. The third step occurred with the evolution of social species. The fourth step appears to be co-incident with the evolution of consciousness.

The development of reasoning ability

The mastery of referential language enables individuals to construct and to describe models of possible situations. These abilities are necessary precursors to the development of reasoning. A system of deductive reasoning also needs to be able to construct models that would falsify an assertion. The system needs to construct and to evaluate such counterexamples, and it needs to be able to do so under the guidance of a particular putative conclusion. Indeed, as I argued earlier, this ability is the essential component that is required to convert a system that is already able to use language into one that is able to reason deductively. Once an individual has the ability to reason, access to a model of the ability enables the individual to think about reasoning at a meta-logical level and to grasp the notion of validity explicitly. There is evidence that literacy also plays a part in the formation of meta-logical knowledge (see Luria, 1977; Scribner, 1977). Writing is, in effect, a way to extend the capacity of working memory and thereby make it possible to carry out algorithms of a greater computational power. It presumably also extends the ability to envisage hypothetical states of affairs. The landmarks in the development of reasoning are summarized in Figure 5.2.

Deductive reasoning depends on an ability to use referential language and to construct models that are counterexamples to putative descriptions. It is therefore clear that children have to master the syntax of their particular language and the meanings of its words, and they have to possess a working memory with sufficient capacity for the search for counterexamples. I will now examine the roles of both of these factors in the development of reasoning.

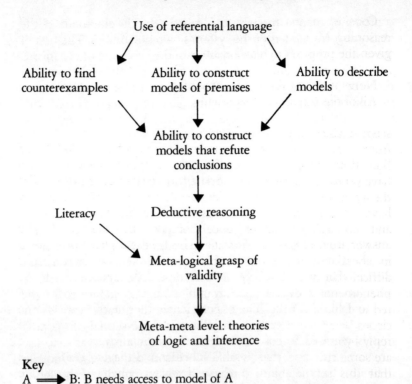

Figure 5.2   Landmarks in the development of deductive reasoning

The dependence of reasoning on an understanding of the meaning of premises may seem to be obvious. The reader should bear in mind, however, that no such prediction is made by a theory that assumes that reasoning depends on a mental logic. Such a logic would enable you to reason without having to know the meanings of premises. As soon as you have attached logical terms such as "none" and "all" to rules of inference, then you would be able to use these rules to make deductions. Just such performance is illustrated by current computer programs for evaluating inferences in the predicate calculus. They reason using one or more formal rules of inference and have no grasp of meaning whatsoever.

Consider a particular case history in the development of reasoning. My colleagues and I have found that 7-year-olds who are given the premises of a syllogism, such as:

None of the footballers are musicians.
All of the runners are musicians.

seem unable to select the correct conclusion with any reliability from a set of only three alternatives (Johnson-Laird, Oakhill and Bull, 1986). One explanation for their incompetence is that they have yet to develop the appropriate formal rules of inference. The theory of mental models provides an alternative explanation: they have yet to master the meanings of such words as "all", "some", and "no". Is there any evidence to support this prediction? The answer, ironically, is that Piaget and his colleagues have provided it in abundance. Children of this age manifest a characteristic difficulty in understanding "all". A typical demonstration of the phenomenon uses, say, an array of blue circles, together with some red and blue squares. The experimenter then asks: "Are all the circles blue?" A 7-year-old typically replies "No" and justifies the reply by asserting either "there are some blue squares" or "there are some red and blue squares". Inhelder and Piaget (1964) found that this sort of error is more frequent when the predicate concerns colour rather than shape. It is still more prevalent when the predicate concerns weight (Lovell, Mitchell and Everett, 1962). Hence, it seems that children grasp the meanings of sentences better when they concern graphic properties of objects. Likewise, Markman and Seibert (1976) have shown that children have a still better grasp of relations between sets when the superordinate set is one that is characteristically relational, e.g. a family or a bunch of grapes. Donaldson (1978, p. 67) reports that the saliency of a particular aspect of the situation may also mislead children.

These findings strongly suggest that children have difficulty in understanding the correct meaning of quantified assertions, and that they have only a tenuous grasp of their truth conditions, which is easily overborne by other extraneous factors. If children do not understand the premises, then they cannot form a proper mental model of them and they will be unable to reason deductively. Once children have learned the meanings of quantifiers, they should be able to reason more proficiently. We have found that 9-year-olds have a reasonable grasp of the force of quantifiers, and they are able to make valid syllogistic inferences, even drawing their own spontaneous conclusions well above chance for those premises that require only one model to be constructed (Johnson-Laird, Oakhill and Bull, 1986). In a study of 11-year-olds, we tested their grasp of the meaning of individual sentences containing quantifiers. Their ability to make syllogistic inferences correlated significantly with their performance in this task (Spearman's rho = 0.61).

Martin Braine and his colleagues, however, have argued to the contrary that children are able to reason before they have acquired the truth conditions of logical terms. Braine and Rumaine (1981) studied children's ability to understand the meaning of "or" and their ability to make inferences based on its occurence in a premise. In a comprehension task, these experimenters made the following requests:

Give me all the green things or give me all the round things

and:

Give me all those things that are either blue or round.

The children treated these two requests in much the same way, handing over all the members of one set, typically the first mentioned, in both cases. Since the majority of adults who were tested behaved in the same way, the results establish only that there is a preferred reading for the ambiguous second sentence. In a more revealing test, the children heard a sentence of the form:

Either there's a horse in the box or there's a dog in the box

in a context where they could see what animal was actually in the box. They were then asked whether the speaker was "right". The percentages of children making a correct response (on the basis of either an inclusive or exclusive interpretation of the disjunction) increased reliably with age, and the youngest group of 5- to 6-year olds were correct on only eighteen per cent of the trials. This task demands a metalinguistic ability, since it calls for a judgment about whether a statement is, in effect, true or false. A third series of tests examined the children's ability to reason with disjunctions. For example, the experimenter said:

Either there's a horse or there's a cow in the box.

Then the assistant looked inside the box and announced:

There's no horse in the box.

Finally, the child was asked:

Is there a cow in the box?

All the children, even the 5-year-olds, did very well in this task.

Braine and Rumaine argue that their results imply that "the meanings of the natural connectives are given by the inferences to which they give rise, not by truth tables". There are several grounds for caution in extending this claim to the conclusion that children acquire formal rules before they grasp the meaning of connectives. First, and this point cannot be emphasized too strongly, the children may not have been reasoning by using formal rules of inference, but by manipulating semantically based mental models. There is no evidence that their inferences were made on the basis of formal schemata. Second, there was no test of the comprehension of simple propositional disjunctions: the experimental conditions, including those calling for inference, all used sentences with quantifiers in them. These tests were either too difficult for adults or else called for metalinguistic ability. It may be that inferential ability develops prior to a metalinguistic grasp of the truth conditions of disjunctions. The results therefore do not refute the prediction derived from the theory of mental models. For that, it would be necessary to show that children could not understand the sentences, e.g. they could not imagine the state of affairs described by them, until after they were able to reason correctly with them. The same sentences need to be used in both a simple comprehension task that does not depend on metalinguistic ability and an inferential task. The tests should also cover a range of different types of logical terms.

If the ability to use language appropriately is manifested only after the ability to reason deductively, then the present hypothesis will be at risk, though it would still be necessary to show that children can reason using purely formal means and without understanding the meaning of quantifiers and connectives. Until such evidence is forthcoming, it is sensible to suppose that the major task in learning to reason is to master the syntax and the meanings of logical terms.

The other major accomplishment that is necessary for reasoning is the ability to construct mental models that refute potential conclusions. The underlying principle depends on the development of referential language and it is likely to be an innate part of its architecture. But there is a subsidiary requirement: an individual needs a working memory that has the capacity to hold in mind several different models whilst the linguistic system examines them to determine what if anything, they have in common. This common element forms the basis of any conclusion that the inferential system draws.

Jane Oakhill and I have examined adult subjects' ability to seek counterexamples, and we found that they are affected by the "cognitive load" of the conclusion to be refuted, i.e. the more complex its description, the less likely the subjects are to search for counterexamples, and the more likely they are to search for confirmatory evidence (see Oakhill and Johnson-Laird, 1985b). Since their grasp of the principle is not at issue, their failure must reflect factors that prevent putting the principle into practice. One such factor is indeed the processing capacity of working memory. Our study of children's syllogistic ability established two relevant boundary conditions. On the one hand, even the 9-year-olds did better than chance with those premises that do not yield a valid conclusion. They therefore must have appreciated the need to search for counterexamples, and they were able to do so provided that they had only to establish that there are two models of the premises with nothing in common. On the other hand, none of the children performed reliably better than chance with multiple model problems that *do* support a valid conclusion. Despite their appreciation of the need to search for counterexamples, they were unable to do so for these problems. The obvious explanation for this failure is the greater load that the problems place on working memory. It is necessary to find what the models have in common and to formulate an accurate description of it – all whilst the models are held in working memory. This evidence, and other findings in the literature (e.g. Case, 1985), imply that a major maturational factor in the development of reasoning is an increase in the processing capacity of working memory.

Conclusions

If the argument of this chapter is correct, the ability to reason deductively has initially nothing whatsoever to do with the construction of some hypothetical mental logic containing formal rules of inference. On the contrary, it depends on an extension of the concept of counterexamples to the ordinary use of referential language. The comprehension of language calls for a model of the relevant state of affairs to be constructed, and the production of language depends on a process of describing the contents of such a representation. Certain sorts of discourse, however, are open to alternative interpretations. A good description derived from such a discourse is one that holds true in any interpretation: it is a valid deduction from the discourse. This theory implies that reasoning ability depends largely on three factors: a mastery of language, the capacity of working memory, and the ability to search for counterexamples.

How do these abilities develop? I have argued that the linguistic system, including a metalinguistic component, unfolds under the control of an innate program (cf. Chomsky, 1965). The program has emerged as a result of neo-Darwinian evolution. Evolution has, however, equipped the mind with neo-Lamarckian learning algorithms. They do not increase the computational power of the mind, but they do increase its logical power. They enable children to learn the meanings of words and, in particular, the meanings of connectives, quantifiers, and relational terms, which play an important part in underpinning deductive inference. Working memory is clearly a fundamental component in mental architecture: the secret of computational power is indeed a memory for the results of intermediate computations. The development in its processing capacity is presumably a matter of maturation. Finally, there is no evidence that the ability to search for counterexamples develops in any striking way. It is evident in the inferences of young children; it remains labile in adults. It, too , is likely to have evolved as part of the fundamental semantic architecture of natural language. It is, however, susceptible to a major consolidation. Once children are able to reason, they have access to a model of their own ability. Under the tutelage of adults, they may begin to construct a more formal account of patterns of inference (cf. Vygotsky, 1962). Formal logic is a systematic syntax for the search for counterexamples.

References

Braine, M. D. S. and Rumaine, B. (1981), "Development of comprehensions of 'or': evidence for a sequence of competencies," *Journal of Experimental Child Psychology*, 31, 46–70.

Braine, M. D. S. and Rumaine, B. (1983), "Logical reasoning," in Flavell, J. and Markman, E. (eds), *Carmichael's Manual of Child Psychology, 4th edition* (vol. 3), New York: Wiley.

Bruner, J. S., Goodnow, J. J. and Austin, G. A. (1956), *A Study of Thinking*, New York: NY Science Editions.

Bryant, P. E., and Trabasso, T. R. (1971), "Transitive inferences and memory in young children," *Nature*, 232, 456–58.

Case, R. (1985), "A developmentally based approach to the problem of instructional design," in Chipman, S. F., Segal, J. W. and Glaser, R. (eds), *Thinking and Learning Skills* (vol. 2), *Research and Open Questions*. Hillsdale, N.J.: Erlbaum.

Case, R., Kurland, D. M. and Goldberg, J. (1982), "Operational efficiency and the growth of short-term memory span," *Journal of Experimental Child Psychology*, 33, 386–404.

Chomsky, N. (1965), *Aspects of the Theory of Syntax*, Cambridge, Mass: MIT Press.

Donaldson, M. (1978), *Children's Minds*, London: Fontana.

Ennis, R. H. (1975), "Children's ability to handle Piaget's propositional logic," *Review of Educational Research*, 45, 1–41.

Ennis, R. H. (1978), "Conceptualization of children's logical competence: Piaget's propositional logic and an alternative proposal," in Siegel, L. S. and Brainerd, C. J. (eds), *Alternatives to Piaget: Critical Essays on the Theory*, New York: Academic Press.

Evans, J. St. B. T., Barston, J. and Pollard, P. (1983), "On the conflict between logic and belief in syllogistic reasoning," *Memory and Cognition*, 11, 295–306.

Falmagne, R. (1980), "The development of logical competence: A psycholinguistic perspective," in Kluwe, R. H. and Spada, M. (eds), *Developmental Models of Thinking*, New York: Academic Press.

Fodor, J. A. (1980), "Fixation of belief and concept acquisition," in

Piattelli-Palmarini, M. (ed.), *Language and Learning: The Debate between Jean Piaget and Noam Chomsky*, Cambridge, Mass.: Harvard University Press.

Fogel, L., Owens, A. and Walsh, M. (1966), *Artificial Intelligence through Simulated Evolution*, New York: Wiley.

Grice, H. P. (1957), "Meaning," *Philosophical Review*, 66, 377–88.

Hitch, G. J. and Halliday, M. S. (1983), "Working memory in children," *Philosophical Transactions of the Royal Society of London, Series B*, 302, 325–40.

Inhelder, B. and Piaget, J. (1958), *The Growth of Logical Thinking from Childhood to Adolescence*, London: Routledge and Kegan Paul.

Inhelder, B. and Piaget, J. (1964), *The Early Growth of Logic in the Child*, New York: Harper.

Johnson-Laird, P. N. (1983), *Mental Models: Towards a Cognitive Science of Language, Inference, and Consciousness*, Cambridge: Cambridge University Press, Cambridge, Mass.: Harvard University Press.

Johnson-Laird, P. N. (1986), "Conditionals and mental models," in Ferguson, C., Reilly, J., ter Meulen, A. and Traugott, E. C. (eds), *On Conditionals*, Cambridge: Cambridge University Press.

Johnson-Laird, P. N. (in press), "Semantic information: a framework for induction," *European Journal of Cognitive Psychology*.

Johnson-Laird, P. N., Oakhill, J. V. and Bull, D. (1986), "Children's syllogistic reasoning," *Quarterly Journal of Experimental Psychology*, 38A, 35–58.

Johnson-Laird, P. N. and Wason, P. C. (1977), *Thinking: Readings in Cognitive Science*, Cambridge: Cambridge University Press.

Kamp, J. A. W. (1980), "A theory of truth and semantic representation." Report of Center of Cognitive Science, University of Texas, Austin.

Lovell, K., Mitchell, B. and Everett, I. R. (1962), "An experimental study of the growth of some logical structures," *British Journal of Psychology*, 53, 175–88.

Luria, A. R. (1977), *The Social History of Cognition*, Cambridge, Mass.: Harvard University Press.

Markman, E. M. and Seibert, J. (1976), "Classes and collections: internal organization and resulting holistic properties," *Cognitive Psychology*, 8, 561–77.

Marr, D. (1982), *Vision: A Computational Investigation in the Human Representation of Visual Information*, San Francisco: Freeman.

Oakhill, J. V. and Johnson-Laird, P. N. (1985a), "The effects of belief on the spontaneous production of syllogistic conclusions," *Quarterly Journal of Experimental Psychology*, 37A, 553–69.

Oakhill, J. V. and Johnson-Laird, P. N. (1985b), "Rationality, memory and the search for counterexamples," *Cognition*, 20, 79–94.

Parsons, C. (1960), "Inhelder and Piaget's 'The growth of logical thinking.' II. A logician's viewpoint," *British Journal of Psychology,* 51, 75–84.

Scribner, S. (1977), "Modes of thinking and ways of speaking: culture and logic reconsidered," in Freedle, R. D. (ed.), *Discourse Production and Comprehension,* Hillsdale, N.J.; Erlbaum. Reprinted in Johnson-Laird and Wason (1977).

Skinner, B. F. (1948), *Walden II,* New York: Macmillan.

Vygotsky, L. S. (1962), *Thought and Language,* Cambridge, Mass: MIT Press.

Wason, P. C. (1968), "On the failure to eliminate hypotheses … a second look," in Wason, P. C. and Johnson-Laird, P. N. (eds), *Thinking and Reasoning,* Hardmondsworth: Penguin. Reprinted in Johnson-Laird and Wason (1977).

Wason, P. C. and Johnson-Laird, P. N. (1972), *The Psychology of Reasoning: Structure and Content,* London: Batsford; Cambridge, Mass.: Harvard University Press.

6 Causal explanations of cognitive development

James Russell

There is no doubt that developmental psychologists want causal theories.

A purely descriptive developmental psychology would be a very dreary prospect – if this simply meant giving a catalogue of what children can and cannot do at different ages. A developmental psychology describing the underlying mental structures at different stages of development would be a good deal more interesting, but, given the abstract nature of such a theory, there is the danger of presenting causal mechanisms at such a high level of generality that they become lost in a fog of theory. A developmental theory which located the causes of mental change squarely in the environment or "social context", whilst entirely neglecting the endogenous factors which turn these causes into effects, would be hopeless for just the same reason that behaviorism is hopeless: it would not satisfy our curiosity about *psychology*. The converse kind of theory which assumed that the child stands to her environment rather as the seed stands to the soil would, *qua* empirical psychology, probably collapse into the same kind of descriptive dreariness as the first kind of theory. Moreover, it seems downright lazy to explain every acquisition, as a point of principle, as the result of "some change going on in there" – pointing to the child's head. Which leaves us with the most sensible, but the most bland, option: interactionism, the position that mental change is the result of interactions between endogenous and environmental factors. A natural reaction to this is: "But of course. Now tell me about the relative contribution of these two factors in this, and this, and this capacity, and explain why the interaction works this way and not some other way." Interactionism is the most sensible approach, but this is because it is more of a starting assumption than an empirical theory.

Mounting causal theories, therefore, is difficult. Not only is it conceptually difficult – the point I have been making – but it is also empirically difficult, as anybody who has been involved in training or intervention studies will testify. What I want to do in this chapter is to argue that some of the difficulty with locating the causes of development can be mitigated if we do two things. First, we should be fairly modest about what should be encompassed by the term "a causal theory", limiting this to accounts of isolatable changes in processing. Second, we should appreciate the heuristic value of theories which are essentially descriptive of the structure of knowledge development. When we are dealing with high-level theorizing, as opposed to explaining particular capacities, the line between a theory as description and a theory as explanation is difficult, if not impossible, to draw. I will finally give two examples of causal explanation, as I intend the term, in cognitive developmental psychology.

Causal explanation in science

It is frequently said to be the mark of a properly scientific theory that it should be expressed in terms of causes and effects. Science is supposed to tell us about the causes of natural phenomena. This assumption has not gone unchallenged, and perhaps the most famous challenge was presented by Bertrand Russell (1912). Before demolishing the "three mutually incompatible definitions" of the term "cause" which J. Mark Baldwin (ironically one of the founding fathers of developmental psychology as well as a philosopher) produced in his *Dictionary*, Russell points out that physics has ceased to look for causes and "the reason why physics has ceased to look for causes is that, in fact, there are no such things. The law of causality, I believe, like much that passes muster among philosophers, is a relic of a bygone age, surviving, like the monarchy, only because it is erroneously supposed to do no harm" (p. 1).

Noting causal regularities, argued Russell, may be the *first* stage of an enquiry, but after this the scientist gets drawn into "a continually wider circle of antecedents recognized as relevant" (p. 8) – an extension of an argument first made by J. S. Mill some time before. The profoundly interactive (not Russell's term) nature of the universe makes it impossible to operate an atomic level of enquiry. Physicists, he pointed out, use the mathematical notion of a *function* to describe the universe, seeking to capture these mutualities of function by differential equations. So, for example, the acceleration of a particle will have a value that is a function of the masses and distances at a given moment of all the bodies in the system.

It hardly needs saying that this spirited dismissal of causality has not itself escaped criticism. Mackie (1974, Chap. 6), for example, fields a number of arguments against Russell's view. It may be – to cite one of them – that the kind of functional relation approach which Russell took as paradigmatic of scientific explanation is, in fact, atypical. Not all natural processes require the same form of explanation as Newtonian gravity. As Mackie puts it:

It is, indeed, very plain that whereas Russell thought that causality was out of date in 1912, causal concepts are, sixty years later, constantly being used in our attempts to understand perception, knowledge, and memory, and to clarify our thought about action, responsibility, legal claims, purpose, and teleology. (p. 154)

Note that the examples are predominantly psychological.

Mackie also points out that functional laws entail the more familiar "cause and effect" type (he gives them the ironic title "neolithic") of laws – about what happens when something is dropped, for example. If functional laws did not have this property – if they were not grounded in good old causal regularity – they would not be much use to us.

However, when Mackie presents his own view of causal regularity he does not return to the traditional, atomic, one-cause-for-one-effect conception. Taking the J. S. Mill argument for multiple causation as his starting point, he argues that effects are typically the result of conjunctions of factors and that *different* sets of conjunctions may produce the same effect. Thus, phenomenon P might be produced by the conjunction of factors ABC or by DGH or by JKL. Indeed there is no reason why the disjunction should even be *finite*: the natural phenomenon of death, for example, has an infinity of causal conditions. The "cause" of a phenomenon, then, is going to be a, possibly infinite, disjunction of sets of conjunctions. And how are we to regard the *elements* of these conjunctions – the As, the Gs, the Js and so forth? Each such factor is *insufficient* by itself but it is a *non-redundant* part of its own conjunction, and that conjunction is itself an *unnecessary* but *sufficient* condition for the phenomenon taking place. Mackie calls this an "inus" condition – an *i*nsufficient, but *n*on-redundant part of an *u*nnecessary but *s*ufficient condition.

Let us anchor this to a developmental example. If we ask questions like "What causes the development of the skill of reading?" we are going to end up with at least a highly complex and possibly even infinite disjunction made up of conjunctions of factors, *qua* inus conditions. Indeed, if Russell is right, we may not even be able to take the matter this far because these inus conditions will all be interacting with one another! But we will stay with Mackie's thesis for the time being and consider how the question about reading could realistically be narrowed. This time we might ask specifically about the causal contribution of improvements in acoustic short-term memory (see Conrad, 1971). What "causal contribution" means here is "contribution as an inus condition". Thus, there are probably many more than one set of cognitive skills ("conjunctions") that are sufficient for reading development, and no one set is necessary. After all, deaf children can learn to read, albeit with difficulty, so they must be employing a conjunction of inus conditions that does not include acoustic short-term memory. Acoustic short-term memory is insufficient (i.e. the child needs umpteen other skills in addition), but non-redundant (i.e. hearing children presumably *do* use the skill) part of an unnecessary (i.e. deaf children can use other conjunctions) but sufficient condition (i.e. that particular conjunction of skills will result in reading competence).

The thesis that I want to go on to develop is that when we talk about "causes of development" we would do well to restrict ourselves to the conception of cause as inus condition. We have not, however, finished with Russell's more radical case against the traditional conception of causal regularity.

Holism and explanation in cognitive psychology

In taking the example of reading in the preceding section I was tacitly assuming that "reading" means the cognitive skill of translating written symbols into sounds. If we do so regard it then the acccount that we shall have to give of its development must be expressed in terms of information-processing – different kinds of memory, sampling information in eye movements, pattern recognition, and so forth. I suggested that we can regard these components of the skill as inus conditions within a broadly causal explanation. But there is another way of regarding the process of reading – a broader sense. Reading, on this view, does not merely result in a translation into sounds, it results in *comprehension*, and for this to happen the child must deploy her real-world knowledge and her capacity for drawing logical and informal inferences (see, for example, Sanford and Garrod, 1981). In short, reading in this broader sense involves the engagement of the child's total cognitive system, and because of this it seems impossible to make a list of component skills in the way we can in the case of reading in the narrow (as translation to sound) sense.

Perhaps all that this tells us is that we should restrict ourselves to causal (inus condition) explanations of reading in the narrow sense. Well and good; but then what kind of theory of the development of *language comprehension* are we going to aim for? Language comprehension is just too global a capacity, it is just too "big" to attack with the language of inus conditions. In attempting to mount causal explanations for it we are faced with something analogous to the "frame problem" in artificial intelligence (Minsky, 1975): where does the explanation *stop*? "What causes language comprehension?" Well, just about everything the child knows, plus something that we may want to call a "language acquisition device". Here are some similarly wrongheaded questions: "What causes the development of reasoning?"; "What causes memory to develop?"; "What causes the child to develop a personality?"

Does this then mean that we should *abandon* all attempts at high-level theorizing about the development of language, reasoning, memory and so forth? Of course not. But what it does mean, I think, is that if we want to theorize about such global capacities we should take Russell's case against causal explanation seriously. For getting drawn into "a continually wider circle of antecedents recognised as relevant" is exactly what happens to us when we try to mount causal theories of capacities like "reasoning". A person could not reason without the concept of an object, without some knowledge of human intentionality, without knowledge of language, without processing systems x, y, and z ... where do we stop? Such thoughts should lead us, I think, to draw a division between two levels of explanation in cognitive development which I will begin by calling the level of *skill* and the level of *knowledge*.

It is tempting to call this the "performance/competence" distinction; but this is a temptation which should be resisted because Chomsky's (1980) conception of language competence is as a modular, *sui generis*, extremely *un*global system. Instead I will elaborate the skills/knowledge distinction within the context of Fodor's (1983) distinction between "input systems" and "central systems". On Fodor's reading of the evidence, we have a number of domain-specific, purpose-built, processing *modules* to handle incoming information. These modules are "vertical" in the sense that they point outward to the world and do their own particular job without regard either to the systems adjacent to them or to the top-down influence of cognition in the general sense. They deal with sensory input bottom-up. David Marr's (1982) work on vision and his conception of cognitive capacity as being decomposable into a number of modules was an obvious influence on this viewpoint. As is well-known, Marr treated vision as a computational "module" with its own *sui generis* principles of operation which only took account of top-down, epistemic factors in the case of ambiguity. Fodor extends this conception to, among other things, syntactic parsing: "I know of *no* convincing evidence," he writes, "that syntactic parsing is ever guided by the subject's appreciation of semantic or 'real world' background" (p. 78, original emphasis). This is an essentially Chomskian view of grammar-as-a-module. So to put the matter in general terms, the input systems result in our being able to do things in spite of what we know or do not know. In Fodor's dramatic example, if our input systems were penetrable by our real-world beliefs and if a dear friend suddenly decided to poke us in the eye one day, we *would not blink*.

The central systems, by contrast, are "horizontal" rather than "vertical". They are the high-level cognitive capacities which *spread across* input domains. What we believe about motor cars, for example, is not located just within the visual, verbal, or tactual domain. These systems, as Fodor puts it, "exploit the information that the input systems provide" (p. 103). "Problem solving" is a good example of a central system; and analogical reasoning is a still better one because, by definition analogical reasoning involves our moving across domains and drawing out high-level relations that did not exist in the input. Fodor does not deny that central systems *could* have their own kind of modularity (Chomsky believes something like this – see below): he just says (p. 105) that there is no evidence either way.

In the light of this, let us return to language comprehension. If the child is ever to understand what we say to her, she is going to have to develop a whole system of beliefs which have some substantial overlap with our beliefs – we would call this her "knowledge system". This is a "central system" form of acquisition so it should not, according to Fodor, be decomposable into watertight processing modules. As I said, Fodor can offer no evidence for this view so what he does is offer an analogy instead: between the development of beliefs ("belief fixation") in individuals and the development of science – which will bring us back to Russell's arguments against causal explanation in science. Firstly, nothing constrains how a belief or a scientific hypothesis must be confirmed. There is nothing inherent in the belief that snow is cold that determines how the belief must be acquired, and there is no one way in which a hypothesis about how snow flakes are formed has to be confirmed. Secondly, the fixation of a belief in an individual or in the scientific community cannot happen without some accommodation taking place within the system of beliefs/hypotheses that are currently being held. Indeed the point which Fodor stresses is that there is no way, in principle, of *excluding the relevance of one belief to any other belief*. Fodor calls this the property of being isotropic and Quinean;[1] and I shall refer to it by the more familiar, though less precise, term "holistic". So we can say that the central systems – what I earlier referred to as "knowledge" – have the property of being holistic. They do not yield up nuggets of cognitive skill to isolate and study – separate chapters in the book of the higher mental processes – because they form a total system. I think we can say that it was consideration of the holistic nature of the physical universe which inspired Russell's skepticism about capturing natural processes within the language of cause and effect. High-level theorizing in (at least) physics and psychology encounters, what I shall refer to as, "the holism problem".

I intend, therefore, to borrow the term "central systems" as a more precise way of referring to what I earlier called "knowledge".

However I do not want to replace my earlier term "skills" by "input systems" because Fodor means something very specific (there are nine criteria for them) and quite contentious by the term. For our purposes it is better to contrast the central systems with "the peripheral systems" – a broader conception within which the input systems are subsumed. The peripheral systems are the systems which we refer to in cognitive psychology when we use the term "information processing", intending something like "the operation of fixed-capacity mechanical systems". The capacity to process information is something that the central systems utilize but which they do not control or have epistemic access to. Memory capacity is a good example – partly because it clarifies the distinction between input systems and peripheral systems. We do not assume modulary when we talk of memory capacity as we do for the input systems: there may indeed be a number of distinct kinds of memory but we do not have to exclude a priori the possibility of there being some general capacities for storing and retrieving information which spread across domains. But is not memory capacity "cognitively penetrable" on Fodor's criteria? It is true that we may be able to apply the central systems to extend memory capacity (i.e. metamemory), but the actual capacity, *qua* raw material and raw processing routines, that we possess at the outset or at any one time is not something over which we have control. Moreover, we can have knowledge of memory phenomena, but of course we do not know how memory works as a mechanism. Finally, the peripheral systems include what we may call "output systems". That is, we can regard the skillful deployment of attentional mechanisms, and indeed all behavioral outputs not penetrable by the central systems, as output systems. Perhaps the advantages of this more liberal construal of what is distinct from the central systems are now evident.

Description and theories of the central systems

Fodor is pessimistic about there ever being a "serious psychology of central cognitive processes" (p. 129); and if he intends the term "serious" to mean "causal" then the pessimism is well justified. What I want to argue in this section is that we should expect such a serious psychology of the central processes – and thus of their *development* – to be essentially descriptive, and that it will not be any the less scientific for being so.

As a first step we need to look at the relation between the terms "description", "explanation" and "cause". In everyday life and in developmental psychology it is usually quite clear to us whether a statement is descriptive or explanatory. But what produces this clarity is the context in which the statement is embedded, not the form of the statement itself. "It fell in the bath" explains why the towel is wet, though it could just as well be a simple description of the event; "Three-year-olds do not appreciate that there can be two mental representations of one object" can either stand as a purely descriptive statement or it can be used to explain why children of that age find appearance–reality questions difficult (Flavell, 1988). In short, the difference between an explanation and a description is *functional* not formal. If this is so then the difference between a description of something and an explanation of something does not consist in the fact that the latter cites causes and the former does not.

That said, it nevertheless seems to be the case that some statements given in explanation have a more distinctively causal "flavor" than others. This causal flavor is most evident when the statement is made in answer to a question about *something going wrong*. For example, when asked why our car broke down we may say, "The carburetor jets got bunged up with muck from the petrol." This description clearly functions as a cause because it, as it were, *selects itself out* as an inus condition for failure from the plethora of inus conditions for success. It is no accident, then, that we can sensibly ask "What caused the breakdown?" because the conditions for the car *working* are bracketed-off, and thus the holism problem avoided. However, it is either wrongheaded or mystical to ask, "What causes the car to work?"

The same principle applies in developmental psychology. Descriptions of what makes this child a bad reader, or a bad concentrator, or bad at sensorimotor coordination function causally as explanations. "What is causing the difficulty with reading?" is a question we can ask because we can seek the inus condition that is failing to be set in a particular conjunction of unnecessary but sufficient conditions for successful reading. Indeed, as I argued before, if we interpret reading "in the narrow sense" to mean "reading as pure print-to-sound conversion", if, that is, we are dealing with reading as the deployment of peripheral systems, then we may even be able to ask "What causes this child's reading ability?" The answer we would expect to receive would be in terms of a list of component skills or peripheral systems – a conjunction of peripheral systems. On the other hand "What causes the child to comprehend language?" is a silly question because it is a question about the central systems and therefore prey to the holism problem.

At the other extreme, requests for explanation will produce answers with the *least* causal flavour when the request is for some high-level, theoretical account of a natural phenomenon. We may seek an explanation for the fact that planets circle suns, or that human life exists on earth, or that children acquire language. Can we not say that the kind of account we require will take the form of a *description of a mechanism?* Of course it is a mechanism within which some processes bring about other processes; but they do so in virtue of principles which we try to describe, perhaps describe mathematically. Even in our car example, although our description of the workings of the internal combustion engine, transmission and so forth can be given in the form of a serial narrative ("the carburetor turns the petrol into a gas which is then exploded . . ."), it is nevertheless a description of a set of interacting processes. At a deeper level we can ask why petrol gas *does* explode. And the answer we get will be a description of the processes at a deeper, molecular level. To say that the molecular processes *causes* the explosion is wrong: the observed explosion and the molecular change are two levels of description of *one* process, the latter does not in any sense cause the former. A description at a deeper level is not a cause.

So what is the moral of all this? It is that a theory of the development of the central systems may look descriptive, but that this does not mean that it is "unscientific". It will not be a tight theory citing processes from which inus conditions can be isolated and extracted, but this should not be a source of worry to us. Ask yourself what has been the greatest influence on cognitive-developmental research since the last war. I think most of us would say, first, Piagetian theory and, second, Chomskian linguistics. And is not Piaget's theory essentially a general description of the processes of phylogeny and ontogeny, within which human ontogeny can be located? The theory is (notoriously) long on description and short on fine-grain causal mechanisms (which is not to say that the Piagetian picture cannot be filled out in these terms – see Sternberg, 1984). It may be fair to say that the "mechanism" of equilibration is hopeless because it explains too much (Bruner, 1959, p. 369). But this has not prevented it from "causing" us to run experiments which without the theory *would never have been run*.

As for Chomsky, I have already noted the paradox that in Fodor's terms the Chomksian Language Acquisition Device has a somewhat "input process" flavor (e.g. a *sui generis* mechanism which is not cognitively penetrable); so calling Chomsky a theorist of the central systems may look somewhat strained. However, Chomsky's theoretical psychology can certainly be regarded as a description of the development of the central systems. He presents the picture of the mind – central systems – as a set of modules ("cognitive", "linguistic" ...) or "mental organs just as specialised and differentiated as those of the body" (1979, p. 83). The language 'faculty' is seen as an isolatable system – indeed even "easy to isolate among the various mental faculties" (p. 46). Well, is this not a "serious" scientific theory of the central systems? Surely it is, but it is a descriptive rather than causal theory, in that, like Piaget's theory, it describes an underlying reality. In Chomsky's scheme inus conditions would be dealt with under "performance".

Two cases of the operation of inus conditions in developmental research

The discussion has led us to the following position: that it is possible to give causal explanations of cognitive developmental changes if by "causal explanation" we mean "an explanation which cites the fulfillment of an inus condition or conditions". I will now argue that this kind of explanation is afforded when we have situations in which the child has a central system capacity which is not being manifested in some circumstances due to peripheral system inadequacies. Or, to put it less formally, sometimes children cannot act in accordance with what they know because they are inefficient processors of information. If processing later improves then the central system capacity will become manifest in new circumstances.

The next assumption to be made is that in presenting tasks to children we can give them different amounts of "environmental support" (to borrow a phrase from Fischer, 1980). If we carefully ensure at each stage of presentation that the child has encoded the information before asking the question, or if we suggest to the child ways of monitoring her own progress, or if we pretrain on similar tasks then we are giving a lot of environmental support. We can assume that the more environmental support we give, the less likely it will be that the child will experience peripheral system failure – because we are helping her to process the information.

This is how the above remarks can be expressed more formally:
Assumption: Task [a] and task [b] tap the same central system capacity and task [b] gives *less* environmental support to the child.

At age [1] they succeed on task [a]
At age [1] they fail on task [b]
At age [2] they succeed on task [b]

So what has happened between age [1] and age [2] is that the peripheral systems, through which a central system capacity was manifested, have ceased to require the kind of environmental support that was given in task [a]. To say which peripheral system or systems (developed in the interim) are responsible is, I suggest, to give a causal explanation of a developmental change from age [1] to age [2] – an explanation in terms of the fulfillment of an inus condition. Note that a causal question is sensible here ("What caused the change?") for much the same reason that it is sensible when we ask for the causes of *failure* (cf. the car breakdown example). For our experiments will have teased out a failure in an otherwise working mechanism (success on task [a] suggests that the central capacity is *there* – but see below for the caveat!) and so the causal account of development points to where the child now *ceases to fail*.

I will now present two examples of such causal explanation. My conclusion will be that although this kind of explanation is possible, it is difficult to be confident that the fulfillment of the crucial inus condition with time is not something in which the central systems are implicated.

According to Bryant and Trabasso (1971), the Piagetian denial that children below about 7 years of age can draw transitive inferences (e.g. A>B; B>C; therefore A>C) may well be a false negative. The children may simply be failing to remember the premises (i.e. A>B and B>C) at the time of questioning (A?C). In their experiments they very neatly demonstrated that when steps are taken to insure retention of the premises (i.e. when more environmental support is given) children as young as 4 years succeed. We encounter much the same kind of data in research with the elderly (Light, Zelinsky and Moore, 1982). So we have...

At age 4 they succeed on tasks which insure encoding of the premises.
At age 4 they fail on tasks which do not insure encoding of the premises.
At age 7 they succeed on tasks which do not insure premise encoding.

There is no holism problem here because experiments have isolated as an inus condition the peripheral system of short-term-memory capacity.

I must now make two important caveats. First, it may happen that success on task [a] with environmental support is not genuine success at all. For we may be giving the child so *much* environmental support that she may "succeed" without really knowing what she is up to (and knowing what you are up to is the mark of the central systems). The child's success may therefore be entirely the result of our appropriately stimulating the peripheral systems to deliver a correct response to which the central systems make no contribution. For example, in a five-item task (ABCDE)[2] quite extensive training is required if the child is to encode the premises accurately. Well, perhaps this training makes the child encode the crucial items B and D in such a way that B is tagged as big and D is tagged as small *without* the child knowing that A to E form a series. Indeed, what can we say about the data produced by Brainerd and Kingma (1984) which show that children may draw the correct inference in some cases without knowledge of the premises? This certainly looks like a case of peripheral system success, because making a transitive inference means, by definition, understanding how particular premises are related to particular infer-ences. Indeed we may even want to argue that it means understand-ing that they are *necessarily* related in this manner (Russell, 1987a).

These are real and genuine worries, but they are only worries, not knock-down arguments against the possibility of unearthing a genuine (central system) grasp of a logical principle by giving younger children environmental support. *Of course*, overdoing the environmental support will result in false positives; just as underdoing it will result in false negatives, as it sometimes appears to be the case in Piaget's experiments. But this is nothing more than a danger to which we have to be alive. Experimental psychology is an art at least as much as it is a technology; and one of the things that makes experimental child psychology so challenging is the need to be alive to this kind of danger. There is no handbook of rules for avoiding it.

The second caveat is against believing that an account of the peripheral change will explain the central system capacity itself. In this case, it is the danger of believing that the memory data provide us with the material for a theory about what knowledge of transitivity *consists in*. Perhaps Trabasso's (e.g. 1977) theory of the development of the transitive inference is a case in point. Trabasso's basic claim is that children come to draw the transitive inference through encoding the premises in a kind of serial representation and then "reading off" the answer to the inference question (B?D) from this. For example, if the materials are colored sticks A>B>C>D>E then the child may encode this as the representation "green blue red yellow pink" plus the tag that the more leftward the items the bigger. Leaving aside the question of whether the evidence really does mean that children encode a serial representation (see Breslow, 1981, for a good case against), we can ask whether this is a theory of making the transitive inference itself – a central process theory.

What it is, in fact, is an hypothesis about an information-processing procedure that might make correct performance possible. But the construction of a mental representation does not – cannot – *constitute* knowledge of anything, certainly not knowledge of a logical principle, for the following reason. No representation can ever come complete with the rules for its *interpretation*. No picture, no diagram, no table, no map, no model, no formula can ever of itself determine how it is to be employed. It was Wittgenstein (1953) who made the point most forcibly. Thus, if it is indeed true that children form serial representations of the kind Trabasso describes, we still have to explain how they interpret them transitively. Some knowledge base – some central system – must inform this interpretation. If there is *no* interpretation, if correct performance is a mechanical non-epistemic process (as perhaps it is in the Brainerd and Kingma, 1984, studies) then the success is, as I argued above, peripheral process success. Indeed, not only must central system capacities inform the use of any mental representation (as the term being used here),[3] it must also inform the *construction* of the representation. For only somebody who already understood transitivity – at some level – would construct a serial representation of the kind just described. If, on the other hand, *the extensive memory training* brought about the representation without central system involvement then we do not have an example of a truly logical capacity.

Before passing on to the next example, I must admit that I am skeptical about this particular causal hypothesis, mainly because of the assumption that the development of short-term-memory capacity is something that happens independently of the central (metamemory) systems, that memory capacity cannot be a function of what we know in some cases (Russell, 1978, 3.2.). Moreover, we know that children often fail to draw the inference when they have remembered the premises (Russell, 1981). That said, this does nothing to undermine the conceptual "legality" of such a causal theory: it is an *empirical* question (1) whether the time-lag in transitivity performance is a peripheral or central system phenomenon, and (2) *what* peripheral system is involved (e.g. memory may not be the important factor, it may be, say, the capacity to form mental images).

My second example concerns children's understanding of tautology. In at least three studies, to my knowledge, it has been reported that children under about eleven or twelve years of age are very likely to interpret non-empirical sentences (i.e. tautologies and contradictions) empirically (Cummins, 1978; Osherson and Markman, 1975; Russell and Haworth, 1987). For example, Osherson and Markman (1975) found that if the experimenter tells the child "Either the [poker] chip in my hand is green or it is not green" the child will say that she has to look in the hand to determine the truth of the statement. I will concentrate on tautology as an example of a non-empirical statement, making a comparison with performance on contradiction at the end. My main aim is to sketch a causal theory of the development parallelling that for transitivity, before illustrating how difficult it is to maintain the "purity" of causal theories in this area.

A study by Braine and Rumaine (1981) showed that if we present a tautology task in the following way the large majority of the 7- to 8-year-old sample will succeed, and indeed that half of these will be able to organize a good justification for their correct answers. Rather than being asked to assess the status of individual sentences, the children are shown a box and two glove puppets. One of the puppets is made to say "I say there is a dog in the box" and the other is made to say "I say there is no dog in the box". The question is then posed "Does one of them have to be right?" To affirm, on good grounds, that one does have to be right is to show an appreciation of tautology.

We have, therefore, a time lag, comparable to that in the transitivity studies, of around 3–4 years. At age [1] they appreciate the fact of tautology with a good deal of environmental support but not without it, and at age [2] they appreciate the fact of tautology without it. Note that this time the environmental support takes the form of dramatizing for the child the fact that a positive and a negative judgment about one fact together exhaust all possibilities. It is as if, to borrow Johnson-Laird's (1983) terminology, we have half-constructed the child's "mental model" for her. Perhaps an analogous kind of environmental support for transitivity would be presenting the materials in such a way that their seriation was being directly encouraged.

A dog in the box

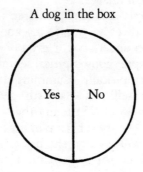

Figure 6.1

What peripheral process capacity might be developing in the interim? As I did for transitivity, I will suggest that we can advance a causal theory of change here in terms of an increase in short-term-memory capacity.

The Osherson and Markman (1975) procedure may be making extra demands on the subject's short-term-memory capacities relative to the Braine and Rumaine (1981) procedure. In the Braine and Rumaine (1981) set-up the child need only construct a model of the following kind; with the circle standing for all possibilities – all possible worlds (see Figure 6.1).

Any possibility must therefore fall in *either* the "yes" half or in the "no" half of the circle. The subject is not being called upon to reflect on a proposition and judge its truth: she must simply appreciate that in this situation, one or other statement will turn out to be true.

By contrast, in the Osherson and Markman (1975) procedure the subject must not only construct some kind of mental model to represent the fact that a possibility must fall in either "half", but must represent this fact *as true* (see Figure 6.2).

What is required, therefore, is a kind of mental quotation of the model in order to make a judgment about it. But is it the mental quotation *per se* that is making the task difficult? This is unlikely, the argument would run, because mental quotation, as Leslie (1988) has recently pointed out, is something that the child would have been doing for at least as long as she has been indulging in pretend play. Perhaps the difficulty is caused by the fact the mental

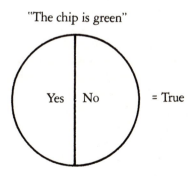

Figure 6.2

quotation is just something extra for the child to do. To illustrate, she must already hold the following two possibilities in working memory.

Either (1) the chip is green
or
        (2) the chip is not green
is true.

But now there is the third step:

        (3) that either (1) or (2) is true *is true*.

Some of Graeme Halford's work supports the suggestion that tasks which require three propositions to be held in mind at one time make particularly heavy processing demands before adolescence (Halford and Wilson, 1980). The basic idea is that in representing a proposition *two* elements have to be considered simultaneously: one of these is the symbolic representation of the fact and the other is a representation of its concrete "environmental" instantiation. Thus, to represent the proposition that A is bigger than B we must represent the A>B relation symbolically whilst also representing one stick, say, being bigger than another. Representing the premises in a transitivity task requires, therefore, *four* elements to be considered (i.e. two symbolic and two environmental). Halford and Wilson (1980) present evidence to the effect that short-term-memory capacity of at least four chunks is necessary for constructing this kind of representation (i.e. is associated with transitive reasoning ability). At the next level of processing capacity ("Level 3"), which they associate with formal operational thinking, three propositions have to be considered (making six elements in all). They provide further evidence that Level 3 tasks require a short-term-memory capacity of at least six chunks.

I present this peripheral process account as an *illustration* of what a causal theory of the development of tautology would look like. In it, short-term-memory capacity of at least six chunks is treated as an inus condition which must be fulfilled if the child is to succeed on tautology tasks of the sentence-assessment variety.

As it happens, I do not find this causal theory particularly satisfying; and the worries are stronger here than in the transitivity case. As before, increasing memory capacity is presented as a cause of development when we could equally regard it as the *effect* of central system changes. One comprehensive study which could decide between these alternatives would be a training study in which we looked at the effect on tautology performance of increasing the (say) digit or letter span memories of children beyond five items.

Another weakness of this causal theory is that it treats the mental quotation of a proposition as if it were just an extra bit of processing, on the assumption that one kind of mental quotation is much like another – be it in sentence assessment or pretend play. This is highly implausible: we cannot equate a sentence and a lego brick as mental objects! It is more sensible to argue, with Osherson and Markman (1975), that treating language as a mental object sets a distinctive cognitive problem for the child and that the status of sentences cannot be properly assessed without this ability. They call this "the objectivity hypothesis".[4]

The third reason for being skeptical about the causal model is based on data rather than on theoretical misgivings. The study of sentence assessment in children between six and ten years which I carried out with Harriet Haworth (Russell and Haworth, 1987) produced the following findings.

1. Not only did children interpret non-empirical sentences empirically, they interpreted empirical sentences (e.g. "The man is wearing red clothes") non-empirically. They often judged these sentences to be true, and their justifications for doing this suggested that they were reasoning that if the experimenter said the sentence or if she presented it to them on a card then it was true. "Because it says so", was a frequent justification.
2. Contradictions (e.g. "The man is wearing blue clothes and he is not wearing blue clothes") were far easier to appreciate than tautologies. Manipulations that improved performance on the empiricals and the contradictions (see note 4) did not affect performance on the tautologies.
3. There was a very strong tendency to interpret sentences containing the word "or" empirically. Thus, the 6-year-olds were significantly better than chance at assessing the

complicated empirical sentence "The man is either wearing red or he is not wearing a hat". I think it is fair to say that this was not genuine "success" at all: rather than "thinking about it" they were hearing the word "or" and immediately saying that they had to look to see if the sentence was true.

4. Performance on the tautologies alone moved from being at chance level in the younger children to being worse than chance in the older ones.

Let us consider the implications of these findings for the causal theory. (1.) suggests a *general* problem with appreciating the status of sentences – a metalinguistic problem – rather than a specific problem with computing one logical form. (2.) means that we have to show why contradiction does *not* involve mental quotation; and it is not at all obvious that it doesn't. (3.) suggests *in conjuction with (2.)*, that the problem with tautology is, at least in part, the result of interpreting any sentence containing "or" as presenting an empirical alternative –*à la* "tea or coffee?". Here, in illustration, is an interchange with an intelligent 12-year-old: Adult – "Tell me if this sentence has to be true, or has to be false, or do you have to look to find out if it's true:" "Either there is a ten pence piece in my hand or there isn't!" Child: "Have to look, because maybe there is one and maybe there isn't, so I need to find out which." (4.) suggests that in the older children the difficulty with tautology is not due to processing difficulties but due to the adoption of an incorrect strategy – see (3.).

To conclude, I hope that this rather chauvinistic excursion into some recent cognitive developmental research has illustrated both the possibility and the difficulty of mounting causal theories in this area. I hope that it has also highlighted the fact that the role causal theories of this kind will play must be decided empirically.

Finally – and with only a hint of irony – I would like to give an example of the kind of difficulty that even intelligent adults have with assessing tautologies. The example is a particularly pertinent one because it concerns a philosophical argument with the conclusion that the development of the central systems can only be regarded as the maturation of an innate system: it cannot be the *acquisition* of knowledge by oganism–environment interaction. (I will be taking "concept acquisition" as being equivalent to central process development because developing concepts is equivalent to developing systems of belief about the extensions of predicates.)

Fodor (1976) has argued that the capacity to acquire a concept entails the prior ability to *represent* the extension of that concept. Therefore, as a point of logic, all concept acquisition must be the unfolding of an innate representational system – a "language of thought". I *suspect* that this argument is so difficult to answer because it is really nothing more than the tautology that "learning X entails the ability to learn X" plus the description of abilities in terms of "representations" – but I am not at all sure.[5]

Notes

1. "Isotropic" because "facts relevant to the confirmation of a scientific hypothesis may be drawn from anywhere is the field of previously established empirical (or of course demonstrative) truths" (*ibid.*, p. 105). "Quinian", after W. V. Quine who, in Fodor's words believes that "the degree of confirmation assigned to any given hypothesis is sensitive to the properties of the entire belief system" (*ibid.*, p. 107).
2. Five items have to be used in order to avoid false positives. For example, with three items A>B>C children may code A as "big" and C as "small" because they are always the bigger and the smaller respectively. They can therefore conclude that A>C without using the middle term. The crucial terms are B?D in the five-item task because neither of them are consistently bigger or smaller.
3. If we use the expression in a more Marr-like sense to mean something like a stage in the neural processing of information then it is not true that a "mental representation" has to be interpreted. See Russell (1987b) for a comparison of the two uses of the term.
4. Our study tested the objectivity hypothesis in two ways. We varied the objectivity of mode of presentation, with written presentation being the most objective and spoken presentation the least. Contrary to the results of Bonitatibus and Flavell (1985), performance was no better with the more objective, written presentation. However, consistent with the objectivity hypothesis, we found that performance improved on the contradictions and on the empiricals (but *not* on the tautologies) when no materials were used – lack of materials may have made it easier to concentrate on the language as opposed to its referents.
5. If we do *not* dismiss the argument as a tautology, what do we say? Perhaps this is one way in which the argument can be defused. Consider concept C which in adults has the extension E. Imagine that the child's first approximation to this concept is C. This is a

protoconcept (Woodfield, 1987) which has no parallel in the adult system. It is wide and amorphous. Indeed the child may begin with one protoconcept whose extension could be, for example, 'the suckable'. Imagine that C_1 has the extension E, F, G, H. How does the child get from C_1 to C? She may do so by moving through a sequence of discriminating E from F, E from G, and so on. Does this imply the prior capacity to represent E? No, it implies the capacity to differentiate between extensions. Does *this* imply the prior capacity to represent the difference between, say, E and F? Perhaps not this either because there is no such thing as *the* difference between two extensions. There are, for example, an infinity of ways in which cats differ from fish products (imagine a child who called the latter "cat" at one time). Discriminating one set of things from another set of things entails the ability to utilize *some* respect in which the sets differ. So far as I can see this does not entail any prior representation of the two sets. To say that it does would be to use the term "represent" with a uselessly wide and amorphous extension.

References

Bonitatibus, G. J. and Flavell, J. H. (1985), "Effect of presenting a message in written form on young children's ability to evaluate its communication adequacy," *Developmental Psychology*, 21, 455–61.

Braine, M. D. S. and Rumaine, B. (1981), "Development of comprehension of 'or': evidence for a sequence of competences," *Journal of Experimental Child Psychology*, 23, 433–76.

Brainerd, C. J. and Kingma, J. (1984), "Do children have to remember to reason? a fuzzy-trace theory of transitivity development," *Developmental Review*, 4, 311–77.

Breslow, L. (1981), "Re-evaluation of the literature on the development of transitive inferences," *Psychological Bulletin*, 89, 325–51.

Bruner, J. S. (1959), "Review of Inhelder and Piaget's 'The growth of logical thinking'," *British Journal of Psychology*, 50, 161–76.

Bryant, P. E. and Trabasso, T. (1971), "Transitive inferences and memory in young children," *Nature*, 232, 456–8.

Chomsky, N. (1979), *Language and Responsibility*, Brighton: Harvester Press.

Chomsky, N. (1980), *Rules and Representations*, Oxford: Blackwell.

Conrad, R. (1971), "The chronology of the development of covert speech in children," *Developmental Psychology*, 5, 398–405.

Cummins, J. (1978), "Language and children's ability to evaluate contradictions and tautologies: a critique of Osherson and Markman's findings," *Child Development*, 49, 895–97.

Fischer, K. W. (1980), "A theory of cognitive development: the control and construction of hierarchies of skills," *Psychological Review*, 87, 477–531.

Flavell, J. H. (1988), "The development of children's knowledge about the mind: from cognitive connections to mental representations," in Astington, J. Harris, P. and Olson, D. (eds), *Developing Theories of Mind*, Cambridge: Cambridge University Press.

Fodor, J. A. (1976), *The Language of Thought*, Brighton: Harvester Press.

Fodor, J. A. (1983), *The Modularity of Mind: An Essay in Faculty Psychology*, Cambridge Mass.: MIT Press (Bradford Books).

Halford, G. S. and Wilson, G. S. (1980), "A category theory approach to cognitive development," *Cognitive Psychology*, 12, 356–411.

Johnson-Laird, P. N. (1983), *Mental Models: Towards a Cognitive Science of Language, Inference, and Consciousness*, Cambridge: Cambridge University Press.

Leslie, A. (1988), "Some implications of pretense for the development of theories of mind," in Astington, J., Harris, P. and Olson, D. (eds), *Developing Theories of Mind*, pp. 19–46, Cambridge: Cambridge University Press.

Light, L. L., Zelinsky, E. M. and Moore, N. (1982), "Adult age differences in reasoning from new information," *Journal of Experimental Psychology: Human Learning and Memory*, 8, 433–47.

Mackie, J. L. (1974), *The Cement of the Universe: A Study of Causation*, Oxford: Oxford University Press.

Marr, D. (1982), *Vision*, San Francisco: Freeman.

Minsky, M. (1975), "A framework for representing knowledge," in Winston, P. (ed.), *The Psychology of Computer Vision*, New York: McGraw-Hill.

Osherson, D. N. and Markman, E. M. (1975), "Language and the ability to evaluate contradictions and tautologies," *Cognition*, 2, 213–26.

Russell, B. (1912), "On the notion of cause," *Proceedings of the Aristotelian Society*, 19, 1–21.

Russell, J. (1978), *The Acquisition of Knowledge*, London: Macmillian.

Russell, J. (1981), "Children's memory for the premises in a transitive measurement task assessed by elicited and spontaneous justifications," *Journal of Experimental Child Psychology*, 31, 300–9.

Russell, J. (1987a), "Rule following, mental models, and the developmental view," in Chapman, M. and Dixon R. (eds), *Meaning and the Growth of Understanding: Wittgenstein's Significance for Developmental Psychology*, Berlin: Springer-Verlag.

Russell, J. (1987b), "Reasons for retaining the view that there is perceptual development in childhood," in Russell, J. (ed.),

Philosophical Perspectives on Developmental Psychology, Oxford: Blackwell.

Russell, J. and Haworth, H. M. (1987), "Perceiving the logical status of sentences," *Cognition*, 27, 73–96.

Sanford, A. J. and Garrod, S. C. (1981), *Understanding Written Language: Explorations in Comprehension Beyond the Sentence*, London: Wiley.

Sternberg, R. J. (ed.) (1984), *Mechanisms of Cognitive Development*, San Francisco: Freeman.

Trabasso, T. (1977), "The role of memory as a system in making transitive inferences," in Kail Jr., R. V. and Hagen, J. W. (eds), *Perspectives on the Development of Memory and Cognition*, Hillsdale, N.J.: Erlbaum.

Woodfield, A. (1987), "On the very idea of acquiring a concept," in Russell, J. (ed.), *Philosophical Perspectives on Developmental Psychology*, Oxford: Blackwell.

Wittgenstein, W. (1953), *Philosophical Investigations*, Oxford: Blackwell.

7 On some relations between the description and the explanation of developmental change

Susan Carey

As this book concerns causes of development – the explanation of developmental change – it begins to redress an imbalance in the developmental literature. Within the field of cognitive development, at least, there has certainly been more research addressing the proper *description* of the changes development brings than research addressing the explanation of those changes. However, this is as it must be. In this chapter I explore two reasons for the priority of descriptive research. The first reason is that we cannot explain developmental change until we know its nature. The second reason, on which I focus, is that the description of the child's conceptual repertoire at any given point constitutes an important part of the explanation of developmental change from that point. I begin with the first reason.

Explanation logically follows description – the case of memory development

The disproportionate emphasis on descriptive research stems from the fact that the explanatory problem cannot even be engaged until we know what developmental changes actually occur, plus the fact that the description of developmental change turns out to be very hard. These points become clear through a very simple example.

A robust finding is that digit span increases markedly between ages four and adulthood; in fact, it more than doubles, from a span of about three at age four to a span of seven plus or minus two in adulthood. But what changes in the child underly this behavioral improvement? One possibility is that the behavioral change reflects a change in immediate memory capacity *per se*; i.e. memory span itself is increasing. With increasing age there are more "slots" in short-term memory. If this is correct, we would be locating the change at a relatively abstract level, at what is called a "domain-general" level, because the change would affect memory for material from any content area or domain. And indeed, memory span for letters, for words (under many conditions), and for many other kinds of materials doubles between ages four and adulthood, which is consistent with the claim that memory capacity itself is changing.

There is, however, another possibility. 4-year-olds, after all, know much less about digits, letters, or words than do adults. Furthermore, adults surely exploit their knowledge of numbers and letters when encoding lists. Perhaps the difference between 4-year-olds and adults is in the knowledge available to organize the input. On this view, 4-year-olds do not have an information-processing limit on memory, do not have a shorter memory span than do adults. They are just less efficient at putting some materials into short-term memory, due to relatively impoverished knowledge of those materials. On this account, the developmental change in question is located at what is called a "domain-specific" level, in the particular knowledge of the materials to be remembered.

To see that more slots and more knowledge are really quite different possibilities, we need only note that it is easy to think of data that could decide between the two accounts. More germaine, it is also clear that the two accounts have quite different implications for possible mechanisms underlying the change. It is difficult to imagine any but a maturational mechanism underlying an increase in the number of slots of short-term memory; alternatively, learning mechanisms clearly underlie acquiring knowledge of numbers or the alphabet. That is, we learn about numbers and the alphabet in the course of such experiences as learning to read and learning arithmetic.

If materials could be found about which 4-year-olds and adults have equal knowledge, the two positions make different predictions. If 4-year-olds have less information processing capacity than adults (the domain-general account) then no matter what the material, their memory span for it should be less. Alternatively, if the domain-specific account is correct, then memory span for materials on which adults and children are truly matched for knowledge should be the same. Chi (1976) reviewed the data on ᵃ᷉e differences in memory span and found two cases for which the ᵊal knowledge condition was met: nonsense figures that are meaningless to both adults and children, and high frequency concrete nouns (like "table," "cat,") which are equally familiar to both adults and children. In both of these cases, the ratio of adult's memory span to 4-year-old's memory span fell from the usual 2:1 to 1.3:1. Equating knowledge of the materials almost erased the developmental differences in memory span.

In sum, it appears that in spite of the robust finding of developmental differences in digit span, letter span, and memory spans for hosts of other materials, *memory span* itself does not change with development, at least after age four. In this case, the most abstract, domain-general level is not the locus of the developmental change.[1]

A distinction must be drawn between two ways in which cognitive development may proceed at a domain–general level. Firstly, the underlying information processing device may change in various ways that has an impact on all computations using that device. Increases in memory span and increases in the speed of the central processor (see note 1) are each examples of domain-general changes of this type. Secondly, knowledge itself varies in its abstractness. Some knowledge, such as mathematical knowledge, has implications for learning and reasoning in many other content domains. Therefore, acquisition of such knowledge would constitute a kind of domain–general developmental change. When seeking to locate the sources of developmental changes, we must consider the possibility of domain–general changes of both sorts.

Consider metamemorial knowledge (knowledge about memory). There is no doubt that young children lack explicit knowledge of many aspects of memory (e.g. they are unrealistic about how much they can remember; they do not realize that some things might be easier to remember than others.) Because they have a relatively impoverished conception of what is involved in memorizing something, they are unable to appreciate the need for mnemonic strategies, and they are unable to devise appropriate ones. For example, they do not cluster items according to conceptual categories for free recall, nor do they rehearse. Indeed, they do not spontaneously use any of a wide range of mnemonic devices that could help them succeed in any given memory task (see Flavell, 1985, or Kail, 1979, for reviews). There is marked improvement in the use of mnenomic strategies between the ages of five and ten. If you teach 5- or 6-year-olds to use a particular mnemonic device (e.g. rehearsal), they can do it, and their performance on memory tasks improves as a result of using it (again supporting the claim that the differences between children and adults on short-term memory tasks is not memory capacity, *per se*). However, they do not spontaneously use this strategy unless specifically instructed to, even after they have been taught it.

Notice that such metamemorial knowledge is relatively domain–general; it applies to the memory of materials of any particular content. The mechanisms by which it is acquired are most likely learning mechanisms; knowledge about memory is part of a rich body of concepts concerning the human mind, concerning knowledge, its acquisition and its use. Throughout childhood the theory of mind is being elaborated (see Astington, Harris and Olson, 1988, for a review); metamemorial development reflects and depends upon this elaboration.

The research reviewed reveals two important sources of developmental differences in performance on memory tasks: (1) domain–specific knowledge of the materials to be remembered, and (2) metamemorial knowledge. The current consensus is that there is little or no development, at least after age four, in memory capacity, *per se*.

This case illustrates the logical priority of the descriptive problem over the explanatory problem. The implications of this priority are not fully drawn until we address a secondary issue, namely, the ease of providing an adequate description of developmental change. Other cases, too complicated to go into in a short chapter, could better illustrate the difficulty of this task. For example many controversies over Piaget's work concern this issue. As in the simple case of short-term memory, the debate concerns whether the locus of developmental change is at the abstract, domain–general level, captured by the stage theory (e.g. the preoperational child is not capable of representing inclusion relations among classes, or of making transitive inferences, while the concrete operational child is) or at a more domain–specific level of particular content domains. Carey (1985a) reviews the controversies concerning this question, and concludes, parallel to the oversimplified example of memory span, that available evidence suggests that Piaget consistently misplaced the locus of developmental change, locating change at the level of domain–general operations when it actually occurs internal to more particular content domains. I will not review these arguments here; my point is simply that these controversies are extremely important, even to those who wish to begin to redress the balance of exclusive concern with the descriptive problem, for we cannot explain development until we have characterized it.

When description is explanation

Students of cognitive development treat developmental change as episodic. A particular developmental change is targeted, the child's knowledge or capacity at time 1 (T1) is assessed, and then the child's knowledge or capacity at a later point (T2) is assessed. The descriptive problem we considered above is how we characterize the *difference* between the child's conceptual system at T1 and T2.[2] The explanatory issue presupposed earlier is the specification of the mechanisms that account for the change from T1 to T2. In that case the relation between description and explanation is one of logical priority. In what remains of this chapter, I will discuss a different relation between explanation and description, that of identity. Describing conceptual structure at T1 simply *is* part of the explanatory enterprise, and a very important part at that.

At any given moment, the conceptual structures people have determine their representations of the world, focus their attention to aspects of the world, constrain the inductions they make over the world, and thus constrain what they learn about the world. That is, the conceptual structure at T1 is a large part of the explanation of the conceptual structure at T2. That the explanatory and descriptive problems collapse in this case is easiest to see when we consider innate knowledge, when T1 is the absolute starting point for knowledge acquisition in some domain. It is perfectly obvious that the characterization of innate knowledge is a terribly difficult part of the explanation of development. However, the characterization of conceptual structure at T1 for any episode of developmental change is equally important, and also difficult, as I will illustrate from an extended example from the literature on lexical development.

I take as my starting point some aspects of the empiricists' proposals concerning the initial state, for I believe these are widely presupposed within psychology today. On the version of the empiricist proposal I consider, humans are endowed with an innate quality space, specified over perceptual features, weighted for salience. Induction is constrained only by similarity in terms of this innate quality space. General mechanism of learning (correlation detection, as in connectionist models, prototype abstraction, and so on) operate over this innate quality space to establish new categories, which, in turn enrich the similarity space over which further inductions operate.

Word learning – T1 of 24 months

Young children are word-learning wizards, acquiring new vocabulary at the prodigious rate of eight to ten items a day (Carey, 1978). Carey and Bartlett (1978) and Heibeck and Markman (1987) have shown that under some circumstances 2- to 4-year-old children posit an initial interpretation of a newly heard word on just one or two encounters with that word used in context, a process dubbed "fast-mapping". Fast-mapping instantiates the classic problem of induction. As such, a central problem for understanding fast-mapping is specifying the constraints on this inductive process. Specifying these constraints amounts to describing the initial state that allows word learning to proceed so efficiently. For the purposes of this example, T1 is just around the second birthday, early into the period in which vocabulary acqusition has begun in earnest. In my discussion of this example, I draw on the work of Soja (1987; Soja, Carey and Spelke, in preparation).

According to the empiricist specification of the conceptual system at T1, newly heard words are projected by the child to new instances of their referents on the basis of a perceptually specified similarity space. For example, Landau, Smith and Jones (1988) suggest that the child's first hypotheses about noun meanings are that the referents of newly heard nouns share common shapes. Similarly, Clark (1973) suggested that early word meaning specified perceptually salient features of the referents the child had heard the word applied to, and Gentner (1978) found evidence that supported this view.

Contrary to the empiricist position, many researchers now believe that in limiting their hypotheses about word meaning, children make many assumptions of many different kinds (cf. Carey, 1982; Clark, 1987; Markman and Hutchinson, 1984; Markman and Wachtel, 1988; Osherson, 1978; but see Nelson, 1988, for a critical opinion). To focus the issue, let us begin with two related sources of constraint suggested by Markman, the taxonomy assumption (Markman and Hutchinson, 1984) and the whole object assumption (Markman and Wachtel, 1988), both of which limit hypotheses about word meaning in the following situation: Suppose the child hears "that's a cup" when the speaker is indicating a brown plastic cup half-filled with coffee. Suppose further that the child knows of no word to refer to any aspect of the situation, so assumptions about mutual exclusivity (Markman and Wachtel, 1988) or contrast (Clark, 1987) cannot help. "Cup" could refer to cups, tableware, brown, plastic, coffee, being half-full, the front side of the cup and the table, the handle, any undetached part of a cup, a temporal-spatial stage of the cup (that is, the particular cup at the particular place and time), the number one, cup-shape, and so on for an infinitude of possibilities. The *whole object assumption* limits hypotheses of the meaning of "cup" to those that include whole cups in the extension of the word. The *taxonomic assumption* limits the hypotheses of the meaning of "cup" to taxonomic categories including cups. Markman and Hutchinson define a taxonomic category as one in which the members are grouped together on the basis of similarity, rather than on the basis of other types of relations (e.g. causal or thematic relations). Together with the principle of contrast, these two assumptions give us the following complex constraint:

(1) If a newly heard word refers to a physical object for which no name is known, that the word picks out a taxonomic category including the whole object.

As stated, constraint (1) is still far too weak to do the work Markman wants of it. Returning to the brown cup: *brown thing, plastic thing*, and *thing I like* are all categories of whole objects defined by similarity to the target object, as is *cup*. We need an analysis that picks out *cup* as the relevant taxonomic category. We will return to this issue below. Also, while the evidence to date is consistent with constraint (1), the evidence is incomplete. As stated, the constraint has two parts, a condition, the "if" clause in

(1), and a consequence. While there is evidence for the contrast part of the condition (reviewed in Clark, 1987), before Soja's studies there was no conclusive evidence that the referent's status as an object plays a role in the constraint.

What could be meant by "status as an object"? Spelke (1985) has provided an analysis of infants' conception of solid objects. Evidence from many sources demonstrates that infants as young as four to six months individuate as single objects those stimuli that are bounded coherent wholes that move in a unitary fashion. Non-solid substances such as liquids, powders and gels differ from objects in all these respects. Otherwise, non-solid substances and solid objects share many properties; any instance of either has a shape, color, texture, etc. While non-solid substances do not meet the criteria that define solid objects for babies, it is not obvious that the distinction is relevant to naming. Newly heard words may be projected to new instances on the basis of a similarity space that is neutral with respect to the solid object/non-solid substance distinction, if, for example, the Landau *et al.* (1988) claim that shape determines noun meaning were correct.

This may seem a strange claim. After all, we already know that children's early common nouns are used to pick out what for adults are taxonomic categories of whole physical objects. (e.g. "doggie," "cracker" "book"; Nelson, 1974; Huttenlocher and Smiley, 1987). But we do not actually know what these words mean to the toddler. These words could refer to shapes in the child's lexicon (e.g. "doggie" means dog-shaped, "cracker" means cracker-shaped, "book" means book-shaped), as Landau *et al.* (1988) suggest. Also, the experimental evidence for the whole object constraint (Markman and Wachtel, 1988), is equally consistent with the hypothesis that the child considers the word to refer to shapes, since the objects of the same kind always shared shapes. It is even possible that the child's meanings might be even more unnatural from the point of view of the adult lexicon. Quine (1969) suggested that early nouns might function most like mass nouns in the child's conceptual system – that is "book" refers to a kind of book stuff, any given book being a piece of book stuff. He suggested that until the child learns the syntax of quantification (determiners, plurals, and quantifiers such as numbers, "some" and "another"), the child's conceptual system does not make the distinction between kinds of objects and kinds of stuff. Quine's proposal is an extreme version of the empiricist position I wish to explore through the present case study.

To address Quine's conjecture, Soja, Spelke and I sought to determine whether the referent's status as a physical object affects the similarity relations that determine the taxonomic category posited by the child. To explore this issue we compared inductive projection of newly heard words when the referents were solid objects with inductive projection of newly heard words when the referents were non-solid substances such as gels and powders. Two constraints are thus being tested, (1) above, and (2):

(2) If a newly heard word refers to a non-solid substance for which no name is known, the word refers to a taxonomic category of substances which includes the referent.

If children can be shown to honor both constraints (1) and (2) before they command the syntax of quantification, Quine's conjecture will be falsified. Accordingly, we tested children at ages 2.0, and obtained speech production samples to assess each child's command of noun phrase syntax. Also, if the child can be shown to honor both constraints, the Landau et al. (1988) hypothesis that early nouns refer to shapes will also be falsified, for shape is irrelevant to taxonomic categories of substances.

To put our question another way: any "taxonomic constraint" must be relative to basic ontological commitments. The potential referents of words must be initially categorized as objects, substances, actions, or whatever the child's ontology allows, before Markman's taxonomic constraint can apply. Thus, Markman's taxonomic constraint already embodies a denial of the empiricist position concerning the initial constraints on word learning. Quine has claimed that it is only through the learning of language that these ontological distinctions are induced; they may constrain conjectures about word meaning in competent speakers who command syntax, but not in presyntactic infants. The data available before Soja's experiments would not help us here, for the taxonomic constraint and the whole object bias have been assessed in quite old preschool children (e.g. in Markman and Wachtel, 1988, the groups of children all averaged $3\frac{1}{2}$ or older; similarly in Markman and Hutchinson, 1984). Consistent with Quine's suggestion, Landau et al. (1988) found that the shape bias became stronger with age, even in the word learning situation. In one of their experiments 2-year-olds did not display it.

Here I will sketch only one of Soja's studies (Soja, 1987; Soja, Carey and Spelke, in preparation). We assessed whether the referent's status as a solid object or as a non-solid substance constrains the child's hypotheses about the meaning of a newly heard word, as in (1) and (2) above, or whether the child's hypotheses are based on a similarity space that is neutral with respect to the solid object/non-solid substance distinction, such that words pick out referents on the basis of shape. To assess Quine's conjecture, we looked at word learning by children of age twenty-four months, and we analyzed their speech production for count/mass syntax. Furthermore, if children command the syntax of quantification and make the syntactic distinction between count nouns and mass nouns, then they may use quantifiers and noun subcategorization as a clue to the meaning of the newly heard noun. Brown (1957) found that 4- and 5-year-old children assume newly heard mass nouns refer to substances and newly heard count nouns refer to objects (see Gathercole, 1986, for a slightly different interpretation). To see whether 2-year-olds make use of this information in our word learning situation, we contrasted two different introducing events: (1) syntax neutral (the new word appeared in neutral quantification contexts, in which both objects and non-solid substances were introduced and referred to as "my xxx, the xxx"), and (2) informative syntax (objects introduced and referred to as "a xxx, another xxx" and non-solid substances introduced and referred to as "some xxx, some more xxx.") Sensitivity to the information in the syntactic context would lead to a greater differentiation of the object trials and the substance trials in the informative syntax condition than in the neutral syntax condition.

Children were introduced to new words by ostensive definitions; the referents were unfamiliar objects made of unfamiliar substances, or unfamiliar non-solid substances. If constraint (1) limits 2-year-olds' inductions about word meaning, when they hear a new word referring to an unfamiliar object they should take it to refer to objects of that kind. We do not know, of course, how 2-year-olds might determine the relevant taxonomic category, even if they were honoring the constraint. For example, "plank" is a taxonomic category of solid objects defined largely by substance.

Therefore, we offered a choice to the child that included one item that did not meet Spelke's test for being an object, and thus could not be an instance of an object of the same taxonomic category as the target. This item was three of four irregular pieces made of the substance of the original. The other item was made of a different substance, but shared the shape of the original (see Figure 7.1; object trial). If the child is honoring constraint (1), his or her choice should be the other object of the same kind, namely the item sharing number, size and shape with the original target. They should reject the choice sharing substance. If constraint (2) limits 2-year-olds' inductions about word meaning, when they hear a new word referring to an unfamiliar non-solid substance, they should take it to refer to substances of that kind. To test this hypothesis, half of the trials were non-solid substance trials (see Figure 7.1). Since non-solid substances do not form bounded, coherent wholes, items containing multiple piles of a substance can be the same kind of substance as the target. That is, if subjects honor constraint (2), they will choose the item sharing substance, ignoring the mismatch in number, size and shape.

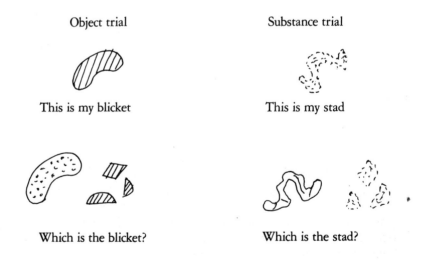

Object trial   Substance trial

This is my blicket   This is my stad

Which is the blicket?   Which is the stad?

Figure 7.1: Sample trials; neutral syntax condition

The details of the procedure and data analyses are presented elsewhere (Soja, 1987). Do note from Figure 7.1 that the non-solid substances were arranged in more distinctive and complex shapes than were the objects (as checked by adult subject ratings). This was to ensure that greater choice based on shape on the object trials would not be due to differential salience of the shapes of the objects. The results are shown in Figure 7.2, which displays the per cent choices based on shape matches minus 50 per cent. Since there were two choices on each trial, chance would be 50 per cent or 0, as displayed on Figure 7.2. Performance differed from chance responding on both the object trials and the non-solid substance trials. When the referent of a newly heard word was a physical object, 2-year-old subjects projected the word to another object of the same kind, respecting the shape and number of the original referent; when the referent was a non-solid substance, 2-year-old subjects projected the word to another sample of the same substance, violating the shape and number of piles of the original referent. As can be seen from Figure 7.2, this tendency was stronger on the object trials (89 per cent correct, overall, when correct is determined relative to constraint (1)) than the non-solid substance trials (61 per cent correct overall, when correct is determined relative to constraint (2)).

As also can be seen from Figure 7.2, there was no difference between the informative syntax and the neutral syntax conditions.

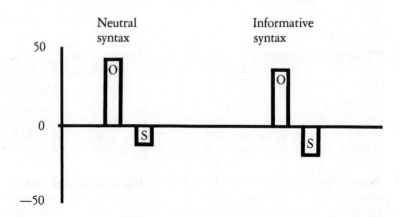

Figure 7.2: Percentage responses matching target in shape and number minus 50

Two-year-olds were at ceiling on the object trials, but far from ceiling on the non-solid substance trials. Nonetheless, hearing the new word in mass noun contexts did not facilitate the induction that it referred to the substance. Also, in their spontaneous production, the children differed greatly in their command of count-mass syntax, from omission of determiners and quantifiers on all nouns to quite a bit of selective use of count noun frames with count nouns. This variation was completely uncorrelated with the individual child's adherence to constraints (1) and (2).

Soja's studies falsify both Quine's and Landau et al.'s (1988) versions of the empiricist characterization of the word-learning system at twenty-four months of age. In what remains of this chapter, I raise several issues concerning constraints on word learning at this age, speculating way beyond what available data allows.

Syntax and constraints on word meanings

These studies falsify Quine's conjecture that the child induces the ontological distinction between objects and substances through learning the syntax of individuation. These data show that the referent's status as an object or a non-solid substance constrains the child's hypotheses about noun meanings at least from age 2:0, well before he has mastered the relevant syntactic devices. An adherent of Quine's conjecture might reply that we may have grossly underestimated the child's command of the relevant syntax, since we have looked only at production. There are now two well-documented demonstrations of syntactic distinctions not under the child's productive control being comprehended, and even constraining initial interpretations of word meaning. Katz, Baker and Macnamara (1974) and Gelman and Taylor (1984) have shown that the 2-year-old is sensitive to the syntactic distinction between proper and common nouns, taking only the latter to refer to taxonomic categories of objects. Similarly, Naigles (1986) has shown that 2-year-olds take into account the syntactic context in which newly heard verbs occur in projections of verb meaning. However, if 2-year-old children comprehend the count/mass distinction, and have induced the object/substance distinction in the course of making sense of that syntactic distinction, then the informative syntax condition should have facilitated their word learning efforts. There was no hint in Soja's studies of such an effect.

We do not know whether syntactic evidence that the new words, "blicket", "stad", etc., were nouns actually influenced the children's interpretations of their meanings, since we included no conditions in which they were not nouns. However, several considerations make it seem likely that the words being nouns played a role in the child's lexical hypotheses. First, that objects and substances are named by nouns is a linguistic universal, and bootstrapping theories of early syntactic categories agree that the noun/object-substance mapping is likely to be innately specified (e.g. Pinker, 1984). Second, the studiest of Katz *et al.* and Gelman and Taylor, mentioned above, show that the syntactic distinction between common and proper nouns constrain hypotheses about new word meaning at this age. Be this as it may, the present studies suggest that within the linguistic category "common noun," at ages 2 and $2\frac{1}{2}$ the syntactic context–mass or count–in which a new noun occurs does not affect the child's hypotheses about its meanings. Rather, the referent's ontological status (in these studies, as an object or as a non-solid substance) seems to determine the child's hypothesis. The correspondence between the conceptual distinction between objects and substances, on the one hand, and the syntactic distinction between count and mass nouns, on the other, is likely learned later (cf. Gordon, 1985; Gathercole, 1986; Brown, 1957).

Ontology and constraints on word meanings

Landau *et al.* (1988) claim that adults, as well as children, ignore ontological categories in their inductive projections of noun meanings. They suggest that for adults, as well as children, noun meanings are determined by shape, irrespective of the ontological categories to which the noun's referents belong. In support of these claims, they offer examples in which the extension of a noun includes objects that share shapes but fall into different ontological categories, in Keil's (1979) sense of different ontological categories. For example, both real bears and stuffed toy bears are called "bear". Or for another example, when asked what an Oldenberg statue of a 100-foot-high clothespin is, people do not say, "a statue of a clothespin"; rather, they reply, "a clothespin". Actually, three different claims must be unpacked here:

1. A noun's extension may include referents of different ontological types.
2. Ontological distinctions are irrelevant to noun meaning.
3. Nouns refer to shape.

We agree with the first claim, but deny the second two.

That a stuffed bear and a real bear are both called "a bear" is just one example of the profligacy with which natural languages unite different ontological types in the extension of single words (see Keil, 1979, for numerous other examples). The lexicons of natural language regularly exploit systematic relations among entities of distinct ontological types. For example, certain geographical entities (countries, states, cities) are the sites of political entities; this relation is exploited in having a single word refer to both, as in "My country is shaped like a boot", and "My country is democratic". Similarly, abstract linguistic objects are instantiated in physical tokens, and we use the same word for both the abstract object and its various types of physical tokens, as in "The sentence was loud", "The sentence was blurred", and "The sentence was elegantly constructed". Stuffed animals, mechanical monkeys and statues are representational objects; the relation of representation is a paradigm example of those which lead to the use of a single lexical item to refer to entities of different ontological types. Thus, while we agree that nouns may well include entities from two or more distinct ontological categories in their extensions, this certainly does not imply that the extensions of nouns are united by sharing a common shape. Countries, as political entities, and sentences, as acoustic or abstract objects, do not have shapes.

Landau *et al.* (1988) may well have meant that when a noun refers to a physical object, shape determines the relevant taxonomic category that unites its extension. But this restatement belies claim 2 above; ontology is of course relevant to constraints on word meanings, since the statement of the claim makes explicit reference to the ontological category of the referent. As Soja's studies show, 2 and 2½-year-old children ignore the shape of non-solid substances in their hypotheses about the meaning of newly heard nouns refering to them. Dickinson (1988) extended these findings to 3- to 5-year-olds and adults. Shape cannot be the *general* taxonomic basis for noun categories, for the simple reason that many types of entities named by nouns, such as substances and abstract entities are not distinguished on the basis of shape. However, it may be that for solid physical objects, and for very young children, shape determines the relevant taxonomic basis for noun meanings. In all experiments to date probing patterns of projection of noun meaning from an initial referent that is a solid physical object, the taxonomically based choices shared shape. However, shape similarity also reflects deeper similarity – similarity of shape follows similarity of parts, and thus similarity of function in the case of artifacts or similarity of evolution and/or genetic forces in the case of biological kinds. We do not know whether the child, even at age two, is after a deeper source of similarity than shape similarity, but we do know that the adult is. Keil's studies of natural kind terms show that adults, and even early elementary aged children, when deciding what a given animal is, are robustly sensitive to how an animal came to get its shape. For example, adults and 10-year-olds are certain that if an antelope were to get a long neck by plastic surgery, it would not become a giraffe, even if the surgeon made it physically indistinguishable from a giraffe (Keil, in press).

As pointed out above, if one defines taxonomic categories simply as those categories determined by similarity relations, as opposed to causal or thematic relations, the taxonomic assumption fails to rule out many hypotheses about possible noun meaning the child would never entertain. What is needed is an analysis that distinguishes cup as a relevant taxonomic category from other similarity-based categories, such as "brown thing" or "thing I like" (or, for adults, even "cup-shaped").

We know of two related proposals for distinguishing a property such as being a cup, which is a good candidate for a property that might determine the extension of a noun, from properties such as being brown or being cup-shaped, which are not. The first is Markman's (in press). She suggests that the properties that are candidates for noun meanings are those with the most inductive depth. If I know that an object is a cup, there are many inferences I can draw about its size, shape, purpose, origin, parts, material it may be made of, and so on. If I know an object is brown, very little follows. Markman suggests that the properties the child entertains as candidates for noun meanings are the inductively deep ones. The second derives from the philosophical literature on metaphysics, in which it is pointed out that to have a concept entails more than to be able to determine its extension. Knowing what a cup is entails being able to trace any given cup through time, being able to know whether one has the same cup on two different occasions, or two different cups. This literature draws (and debates) the distinction between essential and accidental properties. Any property of a given object, such that if that object ceases to have that property, it ceases to exist, is an essential property. Note that essential properties are not concerned with categorization, with what makes something a cup or a dog. Rather, they are concerned with what makes something itself, that is, with individuation and tracing entities through time. Examples of essential properties are those of being a dog or being an animal. If Domino, my pet Labrador, ceased to be a dog or an animal, she would cease to exist (presumably, she would be dead and/or destroyed). However, if she ceased being black, four-legged, in Cambridge, a pet, or having an infinitude of other properties, she would still exist. Perhaps the child's first hypotheses about a newly heard noun is that it picks out a taxonomic category determined by a property essential to the referent.

These two proposals are related in that essential properties are inductively rich properties. The condition of being an essential property is stronger than the condition of having enough inductive depth to be lexicalized as a noun in the adult lexicon. "Pet", "passenger", and many other nouns pick out categories in terms of non-essential properties of their referents. Hall and Waxman (1988) have recently shown that 4- and 5-year-old children's first hypotheses about newly heard nouns that refer to objects for which no label was previously known are that they are taxonomic categories determined by a property essential to the referent, rather than by a temporal stage property such as "being a passenger". They showed their subjects a strange creature climbing a mountain, and told them, "See this, this is a blicket because it is climbing the mountain". They were then asked to indicate another blicket, given a choice of another creature of the same kind in some other context (e.g. riding in a bus) and a different kind of creature climbing a mountain. Subjects choose the new creature of the same kind, ignoring the explicit reference in the introducing event to the fact it was called a blicket because it was climbing the mountain. If however, they already know a name for the creature, they would accept *mountain-climber* as a meaning for "blicket".

We are suggesting here that constraints (1) and (2) should be restated:

1. If a newly heard word refers to a physical object for which no name is known, (a) the word picks out a taxonomic category of objects determined by an inductively rich property of the referent, or (b) the word picks out a taxonomic category of objects determined by an essential property of the referent.
2. If a newly heard word refers to a non-solid substance for which no name is known, (a) the word picks out a taxonomic category of substances determined by an inductively rich property of the referent, or (b) the word picks out a taxonomic category of substances determined by an essential property of the referent.

Just what the essential properties are, or the inductively deep properties are, vary according to the ontological category of the referent. For example, I am the same person I was seven years ago, even though I do not today share a single molecule with myself of seven years ago. Apparently, being made of exactly the same stuff is not critical to being the same person. However, a portion of gold cannot be identical to another portion if they contain no molecules in common. The conditions of identity for substances are different from those for people, whereas the conditions of identity for dogs are deeply similar to those for people. For adults, at least, metaphysical commitments cannot be separated from constraints on word meanings. The present studies suggest this is true also for 2-year-olds, but we do not yet know whether words such as "dog" pick out taxonomic categories determined by essential properties at that age.

Are constraints (1) and (2) language specific?

Constraints (1) and (2) are stated as part of a language acquisition system; they are rules for constraining hypotheses about noun meanings. But it is certainly possible that any induction the child might make would follow these rules. For example, seeing a new object and hearing that if you shake it it rattles, an adult would certainly be more likely to assume that another object of the same kind would rattle than would pieces of that object. After all, we are suggesting that what determines object kind are inductively rich properties. Therefore, other things being equal, any new property might be projected as would candidate word meanings. But for adults, it is not *any* new property. Imagine you encounter an unfamiliar object made of an unfamiliar substance. It has an unfamiliar odor, and it is extremely heavy. In these cases, you would expect an object of a different kind to have the same odor, and also to be heavy, so long as the new object were made of the same substance. Soja (1987), Wiser (1986), and Schmidt (1987) have compared projection of word meaning with projection of odor and weight by children as young as two and a half. These studies show that by age three, weight is projected on the basis of solid substance kind, and by age two and a half odor is projected on this basis. At these ages, with the same stimuli, word meaning is projected according to object kind. That is, young children are sensitive to solid substances, and do make inductive projections based on them. Not *all* inductive projection follows constraint (1). In this sense, constraint (1) is part of the language acquisition device, and not simply part of the similarity space underlying all of the child's inductive inferences. As of yet, we do not know the answer for younger 2-year-olds; we have not found a technique that we can use with toddlers tapping their inductive projection of odor and weight.

What about an earlier T1?

Two-year-old children are well into vocabulary growth. Many researchers claim that word meanings undergo a qualitative change as the child enters the vocabulary spurt at around eighteen months (Dromi, 1986; Nelson, 1988.) This important possibility raises descriptive issues of both types that have concerned us in this chapter. First, if the constraints on word meanings posited by Soja, Markman, Clark and others do not become part of the child's conceptual repertoire until after eighteen to twenty months, what constrains their hypotheses before that time? We must discover a different set of constraints that enable word learning to get off the ground. Second, and relatedly, just what *is* the difference in word meanings at t1 = 12 months and t2 = 24 months?

The data that support the claim of a qualitative reorganization around the vocabulary spurt come from speech production. Indeed, the vocabulary spurt itself is a production phenomenon; we do not know whether there is a vocabulary spurt in comprehension. Thus, we do not know whether the differences reflect changes in the child's productive control of conventions of speech use or changes in the meanings they entertain for newly heard words. The studies of production badly need to be supplemented with studies of comprehension and of word learning before we can tackle these two descriptive problems.

Conclusions

Developmental psychologists properly wish to understand the causes of development. This leads us sometimes to become impatient with the overwhelming preponderance of descriptive research we actually do. My goal in this chapter has been to urge us to be patient. It can be no other way.

Notes

1. This does not mean, of course, that there are no domain–general changes in the information processor. Kail (1986) has recently argued for a maturationally driven increase in the speed of the central processor. Many researchers have noted that children are slower than adults at just about any task on which they are compared. This observation permits two competing explanations, parallel to those for digit span: (1) Novices at any skill are slower than experts; children are novices in every domain; developmental differences in speed may simply reflect domain–specific expertise, due to more practice at every skill. (2) With maturation, the central processor carries out its operations faster. Kail offers two arguments in favor of the second possibility. Using additive factors techniques, he isolated two quite different central processes – deciding whether two tokens of letters, one upper case and one lower case, represent the same letter, and mental rotation – and found identical exponential growth functions for the two types of processes. That is, speed at carrying out each of

these two central processes increased exponentially, and the exponential rate was the same for both. Secondly, he reviewed cases where child *experts* are compared with adult *novices*; children are still slower. These data are suggestive of a maturationally driven change in the speed of the central processor, although this possibility is by no means certainly established.

2. The locus (on the continuum from domain–general to domain-specific) of developmental change is only one component of this descriptive problem. Another, orthogonal, descriptive problem arises within the characterization of knowledge acquisition at any given level of abstractness. It is widely agreed that knowledge is restructured in the course of acquisition, but what, precisely, is meant by "restructured?" See Carey (1985b, 1988) for discussions of this difficult descriptive problem.

References

Astington, J. W., Harris, P. L. and Olson, D. (1988), *Developing Theories of Mind*, Cambridge: Cambridge University Press.

Brown, R. (1957), "Linguistic determinism and the part of speech," *Journal of Abnormal and Social Psychology*, 55, 1–5.

Carey, S. (1978), "The child as word learner," in Halle, M., Bresnan, J. and Miller, G. A. (eds), *Linguistic Theory and Psychological Reality*, Cambridge, Mass: MIT Press.

Carey, S. (1982), "Semantic development: The state of the art," in Wanner, E. and Gleitman, L. R. (eds), *Language Acquisition: The State of the Art*, pp. 347–89, Cambridge: Cambridge University Press.

Carey, S. (1985a), "Are children fundamentally different thinkers and learners from adults?" in Chipman, S. F., Segal, J. W. and Glaser, R. (eds), *Thinking and Learning Skills*, 2, pp. 485–517, Hillsdale N.J.: Erlbaum.

Carey, S. (1985b), *Conceptual Change in Childhood*, Cambridge, Mass: Bradford Books.

Carey, S. (1988), "Conceptual differences between children and adults," *Mind and Language*, 3, 167–82.

Carey, S. and Bartlett, E. (1978), "Acquiring a single new word," *Papers and Reports on Child Language Development*, 15, 17–29.

Chi, M. (1976), "Short-term memory and limitations in children: capacity or processing deficits?" *Memory and Cognition*, 4, 559–72.

Clark, E. V. (1973), "What's in a word? On the child's acquisition of semantics in his first language," in Moore, T. (ed.), *Cognitive Development and the Acquisition of Language*, pp. 65–110, New York: Academic Press.

Clark, E. V. (1987), "The principle of contrast: A constraint on language acquisition," in MacWhinney, B. (ed.), *Mechanisms of Language Acquisition*, pp. 1–33, Hillsdale, N.J.: Erlbaum.

Dickinson, D. K. (1988), "Learning names for materials: factors constraining and limiting hypotheses about word meanings," *Cognitive Development*, 3, 15–36.

Dromi, E. (1986), "The one-word period as a stage in language development: quantitive and qualitative accounts," in Levin, I. (ed.), *Stage and Structure: Reopening the Debate*, pp. 220–45, Norwood, NJ: Ablex.

Flavell, J. H. (1985), *Cognitive Development*, 2nd ed., Englewood Cliffs, NJ: Prentice-Hall.

Gathercole, V. C. (1986), "Evaluating competing linguistic theories with child language data: The case of the mass-count distinction," *Linguistics and Philosophy*, 9, 151–90.

Gelman, S. A. and Taylor, M. (1984), "How two-year-old children interpret proper and common names for unfamiliar objects," *Child Development*, 55, 1535–40.

Gentner, D. (1978), "What looks like a juggy but acts like a zimbo? A study of early word meaning using artificial objects," *Papers and Reports on Child Language Development*, 15, 1–6.

Gordon, P. (1985), "Evaluating the semantic categories hypothesis: The case of the count/mass distinction," *Cognition*, 20, 209–42.

Hall, G. and Waxman, S. (1988), Unpublished data, Harvard University.

Heibeck, T. H. and Markman, E. M. (1987), "Word learning in children: An examination of fast mapping," *Child Development*, 58, 1021–34.

Huttenlocher, J. and Smiley, P. (1987), "Early word meanings: The case of object names," *Cognitive Psychology*, 19, 63–89.

Kail, R. (1979), *The Development of Memory in Children*, San Francisco, CA: W. H. Freeman and Co.

Kail, R. (1986), "Sources of age differences in speed of processing," *Child Development*, 57, 969–87.

Katz, N., Baker, E. and MacNamara, J. (1974), "What's in a name? A study of how children learn common and proper names," *Child Development*, 45, 469–73.

Keil, F. C. (1979), *Semantic and Conceptual Development: An Ontological Perspective*, Cambridge, Mass: Harvard University Press.

Keil, F. (in press), *On the Acquisition of Nominal and Natural Kind Terms*, Cambridge, Mass: MIT Press.

Landau, B., Smith, L. B. and Jones, S. S. (1988), "The importance of shape in early lexical learning," *Cognitive Development*, 3, 299–321.

Markman, E. (in press), *On the Acquisition of Categories and Category Structure*, Cambridge, Mass: MIT Press.

Markman, E. and Hutchinson, J. E. (1984), "Children's sensitivity to constraints on word meaning: taxonomic versus thematic relations," *Cognitive Psychology,* 16, 1–27.

Markman, E. M. and Wachtel, C. A. (1988), "Children's use of mutual exclusivity to constrain the meanings of words," *Cognitive Psychology,* 20, 121–57.

Naigles, L. C. (1986), "Acquiring the components of verb meaning from syntactic evidence," Paper presented at the Boston Child Language Conference, Boston, Mass.

Nelson, K. (1974), "Structure and strategy in learning to talk," *Monograph of the Society for Research in Child Development,* 38.

Nelson, K. (1988), "Constraints on word learning?" *Cognitive Development,* 3, 221–46.

Osherson, D. N. (1978), "Three conditions on conceptual naturalness," *Cognition,* 6, 263, 290.

Pinker, S. (1984), *Language Learnability and Language Development,* Cambridge, Mass: Harvard University Press.

Quine, W. V. (1969), *Ontological Relativity and Other Essays,* New York: Columbia University Press.

Roth, C. (1983), "Factors affecting developmental change in speed of processing," *Journal of Experimental Child Psychology,* 35, 509–28.

Schmidt, H. (1987), Unpublished data, University of Pennsylvania.

Soja, N. (1987), *Constraints on Word Learning,* MIT: Unpublished Ph.D Dissertation.

Spelke, E. S. (1985), "Perception of unity, persistence and identity: Thoughts on infants' conception of objects," in Fox, R. and Mehler, J. (eds), *Neonate Cognition,* pp. 489–97. New Jersey: Erlbaum.

Wiser, M. (1986), Unpublished data, Clark University.

Part IV Causes of Social Development

8 Causes of social development from the perspective of an integrated developmental science

Robert Hinde

Introduction

As we learn more about human development, its study fragments into specialities – social development, cognitive development, emotional development, physical development and so on. This is inevitable but unfortunate – unfortunate because we are dealing with the development of an integrated organism, and the labels social, cognitive, etc. represent pigeon holes that only partially fit nature. However all are influenced by, and influence, the developing child's relationships with family members and others (e.g. Maccoby and Martin, 1983; Vygotsky, 1978; Doise, 1985; Perret-Clermont and Brossard, 1985; Radke-Yarrow and Sherman, 1985). Perhaps recognition of the importance of the network of relationships in which the child is embedded will help to integrate knowledge about development.

Causes?

However it is perhaps not inappropriate to start by making two points about the title of this volume.
First, the term "development", though unavoidable, must be treated with caution. It must not be taken as implying a steady progression directed towards adulthood. A caterpillar is not just busy growing into a butterfly but must first be a good caterpillar, feed efficiently, protect itself from predators, and so on. In the same way the stages of childhood must be seen as delicate adjustments enabling an organism of that degree of development to survive and progress further to best advantage (Tinbergen, 1963).

Second, if the term "causes" implies a possibility of isolating factors, internal or external, that impinge on the individual and push development from stage N to stage N + 1, it is misleading. The development of social behavior (this chapter's primary concern) can be understood only in terms of a continuing dialectic between an active and changing organism and an active and changing environment, with cause and consequence closely interwoven. The most important parts of that environment are the interactions and relationships that the child has with others, and we must start by considering their nature.

Dialectics between levels of social complexity

Consider a child, newly arrived in preschool, interacting with peers and teachers. The course of each interaction depends not only on the child in question, but also on the interactant. In the short term, the behavior of each will be guided by goals for and expectations concerning the course of the interaction. In the longer term, the behavior each can show is determined in part by the interactions or relationships he or she has experienced in the past – and the current interaction will influence future ones. As the child becomes acquainted with particular others, the course of each interaction will be affected by previous ones with the child or teacher in question, and we can speak of them as having formed a relationship with each other. The nature of that relationship will depend upon those of the constituent interactions, and the nature of each interaction will be affected by the participants' evaluations of the relationship (see Figure 8.1). Each of the child's relationships will be influenced by the other relationships he or she forms as the group acquires a structure, and the nature of the structure will depend in part on the characteristics of the component relationships. Each relationship will be affected by social norms and values brought to the situation by the participants, norms and values that are transmitted and transmuted through the agency of dyadic relationships. And these in turn influence, and are influenced by, a "sociocultural structure" of beliefs and institutions concerning in this case, for instance, the nature of the school and its relation to home.

If we are going to understand development we must come to

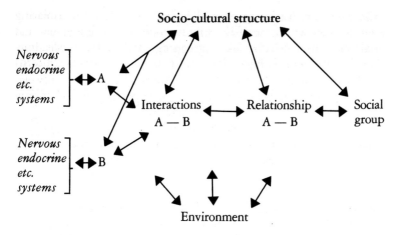

Figure 8.1: The dialectics between successive levels of social complexity.

terms with these successive layers of social complexity, recognizing that each has properties that are simply not relevant to the layer below, that there are continuous dialectical relations between them, and that each is not an entity but a process in continuous creation through the agency of the dialectics. Recognition of these issues provides an essential background for studying social behavior of almost any sort, and especially for studying its development. Whichever level of social complexity one works at, the dialectics almost inevitably obtrude.

Examples of the necessity of seeing a child's behavior as influenced by its relationships will be familiar enough to the developmentalist. When tested in the Ainsworth Strange Situation with their father, 1-year-olds behave differently from when tested with mother (Grossmann *et al.*, 1981; Main and Weston, 1981); the outcome of cognitive tests may depend crucially on the social context and the child's relationship with the experimenter (Perret-Clermont *et al.*, 1984); and preschool children behave differently according to whom they are with (Hinde *et al.*, 1985).

In the present context, individual development can be seen as affected by the higher levels of social complexity in at least three ways. First, for an initially helpless human neonate, interactions with others are essential for survival. Beyond that, through interactions and relationships with others the neonate acquires the skills necessary for living. Second, but related to the last, through relationships with others the child becomes socialized: that is, he becomes progressively less egocentric and able to live in coordination with others. And third, through relationships he becomes acculturated: that is, he learns the forms, models, concepts, schemata, beliefs and ways of interpreting the world that are current in the particular culture in which he or she is immersed. Of course these three types of influence are only heuristically useful abstractions: they proceed concurrently.

In all of these ways, interactions and relationships play a crucial role. As Vygotsky (1978) has emphasized, even the child's discovery of the physical environment is socially mediated, and autonomous thinking arises out of interactions with individuals on whom the child is initially dependent. Development can be understood only with an approch that integrates the effects of different levels of social complexity (Hinde, Perret-Clermont and Stevenson-Hinde, 1985).

Taking account of relationships

We see then that understanding development requires us to treat the child not as an isolated entity but as formed by and forming part of a network of relationships that are crucial to its integrity. This has a number of consequences for the developmental psychologists (Hinde and Stevenson-Hinde, 1987).

The first concerns the way in which generalizations are made from data. Much work in developmental psychology involves studying age trends, or the effects of independent variables, by assessing comparable interactions (e.g. nursing, peer play) across dyads and calculating means or medians. Since interactions of one type may affect interactions of other types within the same relationship, and since relationships may have properties not relevant to isolated interactions, this is not sufficient for full understanding of the genesis of individual differences. Rather we must look at different types of interactions within one relationship, in an attempt to see how the relationship as a whole affects the developing child. This is an ambitious aim, but studies of the correlates of authoritative, authoritarian and permissive parental styles have already shown the value of examining more than one type of interaction simultaneously (Baumrind, 1967, 1971).

That the significance of an interaction of one type may depend upon the incidence of other interactions within the relationship has another consequence for the research worker. In assessing how relationships at one stage effect behavior at another, linear correlational analysis may be misleading. For instance, strong maternal control may not have the opposite sequelae to permissiveness. Other forms of statistical analyses, including categorical discriminant analysis, may yield different insights (Hinde and Dennis, 1986).

Second, we must recognize that many measures of social behavior in fact assess the relationships of the individuals concerned as well as their individual characteristics. Stevenson-Hinde (1985) has suggested that measures in a social situation can be arranged along a continuum from person measures to relationship measures. Height and weight, but few psychological characteristics, are at the person end; the temperament dimensions near, but at varying distances from the person end, and the Ainsworth attachment categories near the relationship end.

Third, since the behavior shown in a social context depends on both (or all) participants, we must seek for ways to tease apart their roles. Various methods are available but they depend on phrasing the questions asked rather precisely. For instance we must distinguish the questions "Do differences between mothers or differences between infants account for differences in the frequencies of control episodes between two mother–infant dyads?" from "Do changes in mothers or changes in infants account for changes in the frequencies of control episodes over a given time interval?" A variety of techniques are available (Hinde and Stevenson-Hinde, 1987).

Fourth, understanding of social development will require us to cross the levels of social complexity, coming to terms with the various dialectics shown in Figure 8.1. Some examples will be given in later sections.

Biological factors: individual differences

Each developmental progression involves a dialectic between the prevailing social situations and the individual as he or she is over a period of time. The latter depends in part on social and in part on biological factors. ("Biological" is used in this context to include genetic influences and also environmental influences that are common to virtually all individuals, but there is no implication that, say, presence or absence of a sibling is not (in some sense) biological). Because the social and other experiential factors vary between individuals, play a large role in the genesis of individual differences, and can be manipulated experimentally, they have become the main concern of the developmental psychologist. Biological factors tend to be neglected. Yet we can obtain only a very partial view of development if we neglect them. To redress that balance it may be appropriate to devote a little space to them here, though without any implications as to their relative importance.

Genetic influences are usually studied by heritability estimates derived from comparisons between mono- and dizygotic twins, or between parents and children brought up in the family of biological origin or in adoptive homes. This has produced an extensive body of knowledge concerned especially with IQ and temperament (Scarr and Kidd, 1983; Buss and Plomin, 1984). These studies cannot be considered further here, but it is in keeping with the theme of this chapter to emphasize that the teasing apart of environmental from genetic influences is by no means an easy task. In general, heritability estimates must be treated with considerable caution (see e.g. Bateson, 1987). More particularly, genetic differences do not merely affect behavior. First, they may affect responsiveness to the environment and predispositions to learn (see below). Second, the child actively selects and creates his or her environment from that which is provided, and how the child does so will be affected by its genetic constitution (Jaspars and Leeuw, 1980). This is termed "active" genotype – environment correlation. Third, parents may be predisposed genetically to give their children both genes and an environment conducive to the development of particular characteristics ("passive" correlation). For example, shy parents might not only pass on "shy" genes but also create an environment in which children see few strangers. Finally, children with different genotypes may call forth different behavior from others ("reactive" correlation) (Plomin and de Fries, 1983; Scarr and McCartney, 1983; Scarr and Kidd, 1983; Buss and Plomin, 1984).

Biological factors: universals

The heritability estimates usually used to study genetic influences cannot detect genetic influences that are ubiquitous: if all variation is environmentally induced heritability will be zero, but of course this does not mean that genes are unimportant. Before proceeding, it is necessary to emphasize that there is no implication here of a return to the old dichotomy between innate and learned behavior. Such a distinction is not only false but sterile (Hinde, 1968; Lehrman, 1953, 1970). Nevertheless it is sometimes useful to categorize items of behavior according to their lability vs. stability in response to environmental influences. Whilst some aspects of behavior are strongly susceptible to environmental influences others are not only pan-cultural, but appear in virtually all environments in which life itself is possible. These must play an important part in shaping the dialectics between the successive layers of social complexity, and must not be disregarded. Consider the following issues:

Ubiquity

Much human behavior is found in virtually all individuals – walking, smiling, sucking, chewing, and so on. Many characteristics of our perceptual systems (e.g. responsiveness to straight lines and circles) are equally ubiquitous. Many human propensities are also common to all individuals – to seek food, to repeat responses leading to food and to avoid responses followed by pain, to be active, to seek to make sense of the environment, and so on. Such aspects of behavior must depend on genes and/or environmental influences common to all individuals.

Development in the absence of supposedly relevant types of experience

Many expressive movements, such as those of smiling, distress, crying, etc., appear in children born blind and deaf, and even in thalidomide children. Learning by example or reinforcement is thus highly improbable (Eibl-Eibesfledt, 1972). That the situations that elicit these expressive movements, the degree to which they are repressed or exaggerated, and their consequences, are subject to subsequent experiential modification is not here relevant (Ekman and Friesen, 1969).

Comparative evidence
Some human expressive movements are so similar to those of non-human primates that a common genetic basis seems probable (e.g. van Hooff, 1972).

Artefacts
Some human artefacts seem to be shaped to satisfy ubiquitous human behavioral propensities. For example, during this century the physiognomy of teddy bears has changed away from being bear-like towards (and beyond) proportions characterizing a human baby (Hinde and Barden, 1985). The latter are believed on circumstantial and experimental grounds to elicit nurturant responses (Lorenz, 1950; Gardner and Wallach, 1965).

Correlations between diverse aspects of human behavior
In some cases diverse human characteristics can be seen as forming an integrated pattern when considered as adaptations to our "environment of evolutionary adaptedness" (Bowlby, 1969). For example many aspects of infant behavior, including the so-called "irrational fears of childhood", many aspects of parental behavior, including differential treatment of first, later and lastborn children, and many aspects of parent–child relationships make sense when seen as adaptations to the conditions under which our species evolved (Hinde, 1984a). Aspects of the ways in which we distribute aggressive and altruistic behavior (Hinde, 1986) and aspects of the differences between the behavior of men and women in close personal relationships (Alexander, 1979; Hinde, 1984b), also form a meaningful pattern when seen as possible adaptations to our environment of evolutionary adaptedness. Although this implies some genetically based differences in learning propensities (see below), it does not imply inherent sex-based predetermined differences in gender role behavior.

Thus a variety of features of human behavior are ubiquitous. This does not of course mean that they are invariant – there is no characteristic not subject to individual variation. Nor does it mean that all aspects of human cultures and cultural differences can be seen as the result of natural selection: the biologists' evolutionary argument, applicable to individual propensities, and illuminating some aspects of interpersonal relationships, is much less valuable when applied to the socio-cultural structure. But just because features are ubiquitous, we tend to neglect them, Yet if we are to understand human development we must come to terms with stable influences and how they interact with experiential ones as well as with the experiential factors themselves (see e.g. Lumsden and Wilson, 1981). Indeed, as shown by the history of schedule feeding, the belief that fear of being left alone is irrational, and the dogma that children must learn to behave as they will have to go on, we neglect these influences at our peril.

Constraints on the predispositions for learning

This emphasis on the role of biological factors is not a form of biological determinism. One of the more important issues thrown up by studies of non-human species in recent years concerns the existence of constraints on or predispositions for learning. To take first an animal case, many song-birds have to be exposed to the species-characteristic song in order subsequently to produce it, but can only learn certain specified song-types. In the chaffinch the constraint seems to be imposed through the note structure – chaffinches learn only songs with a note structure resembling that of normal chaffinch song. In other cases the young birds are more or less restricted to learning songs of the type that the male that reared them sang (Thorpe, 1961; Kroodsma and Miller, 1982). In the same way, constraints and predispositions seem to operate on learning in our own species. For instance, fear of snakes and some other phobias are widespread but not universal (Marks, 1969): apparently a universal propensity can be diminished or exacerbated by experiences in childhood. (This does not mean that snake phobias can be accounted for solely in terms of biological factors. The role of snakes in our mythology, itself to be accounted for in terms of the dialectical relations between the levels of social complexity, no doubt also plays a role.) Some cultural norms and social stereotypes seem to owe their effectiveness to the fact that they harmonize with slight predispositions. We shall consider a specific case shortly.

Perhaps even more important than constraints and predispositions as to what is learned in particular contexts is the fact that humans are equipped to acquire knowledge in specific ways. A baby is predisposed to examine and experiment with some stimuli more than others, and to respond to novel objects in specific ways that will facilitate the learning of their properties. A baby is also predisposed to use social referencing and to act within the relationships he or she has established to relate acts to outcomes, to acquire language as a tool for regulating behavior, and to acquire tool-mediated actions from the pervading culture. Reciprocally, adults are predisposed to provide the relationships that make such advances possible, and to teach the young.

Alternative strategies

To emphasize again that the preceding discussion of biological factors does not imply biological determinism, another issue arising from studies of animals may be mentioned. In studying individual differences, ethologists have found that rather than describing the "normal" behavior of a species with a degree of natural variability around it, it is more profitable to describe the variation as involving alternative strategies for use in different circumstances – the word strategy carrying no implications of conscious planning. Consider the relationships between adult male monkeys. The adult males in a troop can usually be arranged in an approximately linear hierarchy such that A bosses B, B bosses C, and so on. Being at the top of the hierarchy carries a number of advantages in terms of access to mates, food and so on, so that animals strive to be top. But an animal which finds itself low in the hierarchy does not strive continually to defeat those above it. Rather it finds alternative means to satisfy its needs – stealth, subtlety, even deception. Thus we must describe the animal's behavior in terms of a series of alternative strategies – "Be boss if you can, but if you can't, resign yourself and find other routes to your goals." The goals, the range of possible means for achieving them, and the flexibility to adopt one or another, are characteristic of the species: the particular course adopted differs between individuals, and in some cases in the same individual at different times, according to circumstances and the nature of the individual in question. One implication of this view is that, from a biological perspective, there may be no "ideal" style of mothering. Rather each mother must adapt her style according to the situation in which she finds herself.

A comparable argument has been applied to the behavioral styles of human infants with their mothers. In the Ainsworth Strange Situation procedure (Ainsworth *et al.*, 1978) infants are separated from the accompanying parent for about three minutes. On reunion, most infants go straight to the parent and make physical contact with him or her before continuing to play with the toys provided. However, some infants actively avoid and ignore the parent even when he or she seeks their attention, and focus on toys and the inanimate environment instead. In view of the importance of the parent–child relationship to the infant, such behavior seems maladaptive. However, it has been found that the mothers of such children often have an apparent aversion to physical contact with their child and are restricted in emotional expression (Main and Stadtman, 1981). In addition, the child's avoidance is often associated with traces of maternal "anger" (George and Main, 1979). Recently Main and Weston (1982) have suggested that avoidance is a strategy that permits the offspring of such mothers to maintain organization, control and flexibility in behavior. If the infant were to attempt to cling to the somewhat rejecting mother, distress and behavioral disorganization would be likely. The child thus employs alternative strategies according to the nature of the mother.

Sensitive periods

Just because the individual is changing, the environment is changing, and the individual's perception of and sensitivity to the environment is changing, we must expect particular environmental influences to be more effective at some stages in the life cycle than at others. Song learning in birds, described above, provides a classic example: details of the species characteristic song are acquired by example only if the male chaffinch has heard that song during an early period in its life. Similar examples occur in human development. Learning a language and, particularly, learning to articulate its distinctive sounds, is much easier in childhood than in later life (Lenneberg, 1967). This is not the same as saying that there are critical periods for acquiring certain types of experience, after which any deficit is irremediable. The implication is only that there may be phases of greater susceptibility preceded and followed by lower sensitivity, with relatively gradual transitions. The factors behind changes are diverse, and have been discussed elsewhere (Bateson and Hinde, 1987).

The ontogeny of gender differences

The interplay between biological and social factors that lead to gender differences in behavior exemplifies many of the issues discussed in this chapter and may be surveyed briefly.

Young rhesus monkey males show more rough play, threat, aggression and sexual mounting than do females. These differences depend in part on pre-natal hormones; genetic females exposed while *in utero* to exogenous male hormones show the above patterns almost as frequently as do males (Goy, 1978). However these sex differences are affected also by the social conditions of rearing. For example: (1) male rhesus reared in restricted environments show deficiencies in mounting behavior that last into adulthood; (2) infants reared by the mother subsequently show competent sex behavior and low levels of aggression, but infants reared in peer groups tend to be aggressive; (3) the mounting frequency of males is higher in mother-reared heterosexual groups (i.e. containing both young males and females) than in isosexual ones, whilst the opposite is true for females; (4) males reared in heterosexual groups show more aggressive behavior and less fear behavior than those reared in isosexual groups: this is probably because the males occupy only the dominant positions in the former case, whilst in isosexual groups they may be either dominant or subordinate; (5) in contrast to males, even high ranking isosexually reared females show little attack behavior: rearing females away from dominating males does not augment attack or threat behavior; (6) females show less rough play in heterosexual groups than do males, but even less in isosexual groups: this sex difference in heterosexual groups therefore does not depend on suppression by dominant males (Goldfoot and Wallen, 1978).

These results show clearly that, even in monkeys, sexual behavior that is dimorphic in one situation may not be so in another. Social experience clearly contributes to the acquisition of sex-typical behavior. However the authors emphasized that no rearing condition has been found that unambiguously augmented rough play or attack behavior by infant females.

In humans the determinants of gender-role are of course even more complex. As in monkeys, there are genetically based differences operating at least in part through pre-natal hormonal factors and influencing both structural characters and behavioral propensities (Money and Ehrhardt, 1972; Meyer-Bahlburg, 1984). Social experience, however, again plays a crucial part, but in a more complex way than in monkeys. There are differences first in the way mothers and especially fathers treat boys and girls (Tauber, 1979; Jacklin, Di Pietro and Maccoby, 1984). Parents also differentially reward sex appropriate behaviors (Tauber, 1979; Langlois and Downs, 1980). Parent–child relationships may be affected by child characteristics differently in boys and girls: thus Simpson and Stevenson-Hinde (1985, see also Hinde, Stevenson-Hinde and Tamplin, 1985) found that shy boys had more tense home relationships than non-shy boys, whilst for girls the opposite was the case.

In addition, or as a consequence, boys and girls differ in the ideas they acquire about behavior appropriate for the gender they conceive of themselves as having. During the last three decades a large number of studies have documented the existence and effectiveness of social stereotypes in determining gender role (e.g. Block, 1983; Tauber, 1979): many have considerable cross-cultural generality (e.g. Williams and Best, 1982).

It seems likely that such stereotypes are based on very small differences in developmental propensities between the sexes which are greatly exaggerated in the stereotypes for the benefit of one sex or the other. Individuals seek to define themselves in terms of the world in which they live, and one important means to that end involves social categorization, with a tendency to perceive members of the in-group (in this case, own sex) favorably and to disparage the out-group (or other sex). It may also be advantageous to individuals to exaggerate the characteristics that differentiate the groups (Tajfel, 1978; Williams and Giles, 1978). As another example, the large differences between boys and girls in achievement in physics and mathematics found at present in the United Kingdom[1] are undoubtedly due primarily to sex stereotypes which portray girls as poor at mathematics and result in sex-biased teaching, poor gender role models, poor self-expectations etc. But those stereotypes, which are grossly unfair to individuals, may be based on very small differences in mean potential, as exemplified by spatial vizualization (e.g. Scarr and Kidd, 1983). In the same way the markedly divergent stereotypes about how men and women behave in close personal relationships may be based on very small inherent differences in learning predispositions (Hinde 1984b).

Interaction, continuities and discontinuities

At this point it is appropriate to emphasize how some of the strands set out in previous sections are interwoven in development. Development involves a continuing interaction between the individual and his or her environment from conception to death such that each shapes the other. Just as a canary selects a nest site and builds a nest by weaving together grass and feathers, so we select our environments and build our worlds both literally and metaphorically. Just as a canary's behavioral propensities are enhanced, guided and subsequently reduced by stimuli from the half-completed nest in such a manner that the nest reaches completion, so does our perceived world interact with our propensities to edge us nearer to the completed conceptual world ever just beyond our grasp. Just as the stimuli from the half-completed nest induce endocrine changes that alter the sensitivity of the canary's breast to the structure it has built, so does our experience of the world we build alter our perceptions of and sensitivity to it. Consequences are simultaneously causes in a continuing interchange.

The canary's nest is, of course, only a very partial analogy, because the most important elements in a person's world are the relationships built with others who are also active, striving and changing individuals. The dialectics of Figure 8.1 persist over time.

Put in that way, the course of development seems like a continuum, with each experience affecting the individual to a greater or lesser extent and thus having potential repercussions throughout life. Some experiences do indeed produce lasting effects, and some individual characteristics persist over long periods of the life-cycle. For instance, the rank orders of children on various temperament scales show significant consistency from toddlerhood through early childhood (Bates, 1987); children with problems when 3 years old tend to show deviant behavior five years later (Richman *et al.*, 1982); temperament assessed along a "difficult" to "easy" scale at 3–5 years shows significant agreement with a comparable scale at 18–22 years (Thomas and Chess, 1982); and Mussen *et al.* (1980) found that women showed significant stability on five out of five cognitive variables and ten out of sixteen personality variables from age thirty to seventy.

Of course in some cases such stability can be ascribed to persisting environmental influences. The long-term effects of parental loss may be associated with continued inadequate parental care or institutional unbringing (Brown *et al.*, 1986; Quinton and Rutter, 1986), and the long-term effects of intervention programmes in the preschool years may depend on continued support from parents and teachers (e.g. Clarke-Stewart and Fein, 1983; Lazar and Darlington, 1982). However in other cases, such as the long-term effects of institutional upbringing on the subsequent ability to form close relationships, stable long term changes in the individual seem to be involved (Rutter, 1987).

However there sometimes seem to be marked discontinuities in development. An obvious biological case is metamorphosis: the changes from caterpillar to pupa and from pupa to butterfly involve a dramatic change in structure and behavior. In mammals the change from feeding by sucking to eating solid food may involve a switch to a quite different control mechanism (Hall and Williams, 1983).

In such cases it might appear that experiences before the given change would have no long term significance: once pupated, the pupa seems to "start again" and the tissues reorganize to generate a butterfly. Less dramatic changes have been postulated in human development. For instance, the scores of attentiveness, vocalization and smiling in 2-month-old infants fail to predict their relative scores on the same measures at four months (Kagan, 1978), and a number of temporally concordant changes (increased attentiveness, inhibition, object permanence, distress in response to strangers and to separation) have been taken as evidence for a discontinuity in development towards the end of the first year (Kagan, 1980).

That the rate of development (however measured) is not constant and that some characteristics change much more rapidly during certain periods (e.g. puberty) than others, is not to be denied. But great care is necessary before marked discontinuities in development are postulated. There are a number of reasons for believing that consistent causal chains persist through the individual's life, though they may not be immediately apparent and may involve changes in overt behavior appearing to imply inconsistency (Hinde and Bateson, 1984c).

In the first place, the effects of experience may persist through marked structural change. For instance, the food plant selected for egg-laying by certain moths may be affected by experiences they had as a caterpillar (Thorpe and Jones, 1937; Manning, 1967), and the effects of training larval amphibians can be detected in adulthood (Herschkowitz and Samuel, 1973).

Second, the evidence for discontinuities is often misleading. Two examples will suffice. A discontinuity is often postulated on the grounds that the rank order of individuals on a particular characteristic changes dramatically from one time to another. However it could be that one or other set of measurements was made during a period of rapid change, which individuals pass through at different rates, the original order being restored thereafter. Or it could be that in one or other period the characteristic in question was at a floor or ceiling level (e.g. all individuals were able to read perfectly), so that rank order correlations lost their meaning (Bateson, 1976).

Third, many aspects of a child's behavior depend on circumstances that seem at first sight not immediately relevant to the test situation. In the short term, as we have seen, behavior in a social situation may depend on the other participants, and performance in a test situation may depend upon the child's relationship with the experimenter. In the longer term, a child's behavior may be much influenced by other circumstances of its life: for instance, every preschool teacher knows that the appearance of problems in school may reflect trouble at home. Although it appears that parents behave with considerable consistency to successive children of a particular age (Dunn, Plomin and Nettles, 1985), parental behavior often changes dramatically as the child grows up (Clarke-Stewart and Hevey, 1981) so that at any one time parents may treat children of different ages very differently (Dunn, Plomin and Daniels, 1986). Thus just as continuities may be due to consistencies in a child's relationships, so discontinuities may be due to changes in a child's relationships or other background factors in his or her life.

Finally and most importantly, the multiple interacting factors that influence the individual in development may make it exceedingly difficult to identify readily the long-term effects of experience on development. Continuities over time (or consistencies at any one time) may be revealed in ways other than by comparing similar patterns of behavior over time. What we must search for has been termed by Sroufe (1979) "coherence across transformations" or causal connections between experiences at one age and subsequent psychological or behavioral outcomes. At one level, this involves recognizing that the same behavioral propensity may be revealed in different ways at different ages or in different situations. At another, it means that we must come to terms with the ways in which one type of experience or relationship may affect behavior of a different sort in a different context and/or at a different stage in the life history. As an example in a study of the effects of institutional rearing, Rutter, Quinton and Liddle (1983) found that the strongest effect on women's parental styles was provided by the characteristics of the women's spouses. This, however, could be seen as an indirect effect of the institutional rearing: over half of the institution-reared women married men with psychosocial problems, as compared with thirteen per cent in the general population comparison group. Thus the childhood experiences influenced the choice of spouse, and the spouse affected the quality of parenting. A difficult childhood can enhance the risk of psychosocial difficulties in adulthood even though none are apparent at the time (Quinton and Rutter, 1986).

Another interesting example is provided by the work of Sroufe *et al.* (1985) on "generational boundary dissolution" between mothers and their children. Some mothers behave "seductively" (i.e. with certain kinds of physical contact, sensual teasing, etc.) to their children, especially to boys. Sroufe and colleagues showed that this was linked with a reported history of emotional exploitation by the mother's own fathers, that it was stable over the 24–42 month age period, and was often accompanied by a particular type of hostility ("derision") to daughters. They comment that the mother's "Feelings of low self-esteem and emotional need may be manifest with boys in one way and girls in another, depending on her role-relations history, but she is the same person acting with regard to the same history".

This example relates the issue of long-term continuity to that of cross-situational consistency. As we have seen, even preschool children behave differently with different others (Hinde, Titmus, Easton and Tamplin, 1985). In a comparison between the behavior of preschoolers at home and at school, Hinde and Tamplin (1983) found practically no significant correlations in children's ranks on behaviors that could be labelled similarly in home and school (e.g. active hostile to mother and active hostile to peers in school). Nevertheless, there were meaningful patterns of correlations between behavior in the two situations. For instance, children who engaged in few joint activities with their mothers, and exchanged little neutral conversation with them, tended to interact frequently with peers in school. The temperamental characteristic "moody" was correlated negatively with joint activities with the mother at home, but positively with interacting with peers in preschool. Thus as a first step towards understanding the psychological processes underlying the apparent inconsistency in behavior between home and school, we can point to the finding that the temperamental characteristic "moody" seems to be expressed differently in the two situations. Again, the temperamental characteristic "active" was associated with infrequent interactions with the mother at home, and strongly positively with the proportion of interactions with peers in school that involved reactive hostility and similarly for interactions with teachers. Not surprisingly, frequency of interactions with the mother were correlated negatively with reactive hostility in school.

In general, because every behavioral act depends upon numerous capacities, motivational propensities and aspects of the current environmental context, the effects of a given experience on behavior in either the short-term or the long-term are not always easily predictable. This is not merely an attempt to seek refuge from the task of understanding behavioral development in an appeal to complexity. Continuities are sometimes obvious, but when they are not their apparent absence is not in itself a reason for failing to seek understanding of the causal sequences underlying the changes that occur. Although reorganizations of psychological structure do occur, underlying continuities are usually there to be found.

Conclusion

In summary, the developing child must be seen as a social being, and the developmental psychologist must take account of successive levels of social complexity and of the dialectical relations between them. This has implications for the analysis of data about social behavior, implies that measures of social behavior may depend on both participants, and imposes an obligation to tease apart their roles. In considering the nature of the dialectics between levels of social complexity, biological contributions to the nature of individuals must not be neglected. Studies by ethologists and comparative psychologists have some lessons for the study of child development, including the existence of constraints on and predispositions for learning, the occurence of alternative strategies, and the existence of sensitive periods. Finally, understanding of development will come only through an understanding of process.

Acknowledgments

This paper was prepared with support from the Medical Research Council, and the Royal Society. Many of the issues mentioned arose from discussions with colleagues, especially Joan Stevenson-Hinde.

Note

1. See (a) *Girls and Physics*, a report of the joint Physics Education Committee of the Royal Society and the Institute of Physics, 1982; (b) *Girls and Mathematics*, a report of the Institute of Mathematics, The Royal Society, 1985.

References

Ainsworth, M. D. S., Blehar, M. C., Waters, E. and Wall, S. (1978), *Patterns of Attachment*, Hillsdale, N.J.: Erlbaum.

Alexander, R. D. (1974), "The evolution of social behavior," *Ann. Rev. Ecol. & Systematics*, 5, 325–83.

Alexander, R. D. (1979), *Darwinism and Human Affairs*, Seattle: University of Washington Press.

Bates, J. E. (1987), "Temperament in infancy," in Osofsky, J. D. (ed.), *Handbook of Infant Development*, 2nd edn, pp. 1101–49, New York: Wiley.

Bateson, P. P. G. (1976), "Rules and reciprocity in behavioural development," in Bateson, P. P. G. and Hinde, R. A. (eds), *Growing Points in Ethology*, pp. 401–22, Cambridge: Cambridge University Press.

Bateson, P. P. G. (1987), "Biological approaches to the study of behavioural development," *International Journal of Behavioral Development*, 10 (1) 1–22.

Bateson, P. and Hinde, R. A. (1987), "Developmental changes in sensitivity to experience," in Bornstein, M. H. (ed.), *Sensitive Periods in Development*, Hillsdale, N.J.: Erlbaum.

Baumrind, D. (1967), "Child care practices anteceding 3 patterns of preschool behavior," *Genetic Psychology Monographs*, 75, 43–88.

Baumrind, D. (1971), "Current patterns of parental authority," *Developmental Psychology Monograph*, 4 (1, pt. 2).

Block, J. N. (1983), "Differential premises arising from differential socialization of the sexes: some conjectures," *Child Development*, 54, 1335–54.

Bowlby, J. (1969), *Attachment and Loss*, New York: Basic Books.

Brown, G. W., Harris, T. O. and Bifulco, A. (1986), "The long-term effects of early loss of parent," in Rutter, M., Izard, C. E. and Read, P. B. (eds), *Depression in Young People*, New York: Guilford Press.

Buss, A. H. and Plomin, R. (1984), *Temperament: early developing personality traits*, Hillsdale, N.J.: Erlbaum.

Clarke-Stewart, K. A. and Fein, G. G. (1983), "Early childhood programs," in Haith, M. M. and Campos, J. J. (eds), *Infancy and Developmental Psychobiology (vol. 2), Mussen's Handbook of Child Psychology (4th edn)*. pp. 917–99, New York: Wiley.

Clarke-Stewart, K. A. and Hevey, C. M. (1981), "Longitudinal relations in repeated observations of mother-child interaction from one to two and a half years," *Developmental Psychology*, 17, 127–45.

Doise, W. (1985), "Social regulations in cognitive development," in

Hinde, R. A., Perret-Clermont, A-N. and Stevenson-Hinde, J. (eds), *Social Relationships and Cognitive Development*, pp. 294–308, Oxford: Clarendon Press.

Dunn, J. F., Plomin, R. and Daniels, D. (1986), "Consistency and change in mothers' behavior towards young siblings," *Child Development*, 57 (2), 348–56.

Dunn, J. F., Plomin, R. and Nettles, M. (1985), "Consistency of mother's behavior towards infant siblings," *Developmental Psychology*, 21 (6), 1188–95.

Eibl-Eibesfeldt, I. (1972), "Similarities and differences between cultures in expressive movements," in Hinde, R. A. (ed.), *Non-Verbal Communication*, pp. 297–314, Cambridge: Cambridge University Press.

Ekman, P. and Friesen, W. V. (1969), "The repertoire of non-verbal behavior: categories, origins, usage and coding," *Semiotica*, 1, 49–98.

Gardner, B. T. and Wallach, L. (1965), "Shapes of figures identified as a baby's head," *Perceptual & Motor Skills*, 20, 135–42.

George, C. and Main, M. (1979), "Social interactions of young abused children: approach, avoidance and aggression," *Child Development*, 50, 306–18.

Goldfoot, D. A. and Wallen, K. (1978), "Development of gender role behaviors in heterosexual and isosexual groups of infant rhesus monkeys," in Chivers, D. J. and Herbert, J. (eds), *Recent Advances in Primatology*, London: Academic Press.

Goy, R. W. (1978), "Development of play and mounting behavior in female rhesus virilized prenatally with esters of testosterone or dihydrotestosterone," in Chivers, D. J. and Herbert, J. (eds), *Recent Advances in Primatology*, London: Academic Press.

Grossman, K. E., Grossman, K., Huber, F. and Wartner, U. (1981), "German children's behaviour towards their mothers at 12 months and their fathers at 18 months in Ainsworth's Strange Situation," *International Journal of Behavioral Development*, 4, 157–81.

Hall, W. G. and Williams, C. L. (1983), "Suckling isn't feeding, or is it? A search for development continuities," *Advances in the Study of Behavior*, 13, 219–54.

Herschkowitz, M. and Samuel, D. (1973), "The retention of learning during metamorphosis of the crested newt (Triturus cristatus)," *Animal Behaviour*, 21, 83–5.

Hinde, R. A. (1968), "Dichotomies in the study of development," in Thoday, E. and Parker, A. J. (eds), *Genetic and Environmental Influences on Behaviour*, Edinburgh: Oliver and Boyd.

Hinde, R. A. (1984a), "Biological bases of mother-child relationship," in Call, J. D., Galenson, E. and Tyson, R. L. (eds), *Frontiers of Infant Psychiatry*, (vol. II), pp. 284–94, New York: Basic Books.

Hinde, R. A. (1984b), "Why do the sexes behave differently in close personal relationships?" *Journal of Social and Personal Relationships,* I, 471–501: London: Sage.

Hinde, R. A. (1986), "Some implications of evolutionary theory and comparative data for the study of human prosocial and aggressive behaviour," in Olweus, D., Block, J. and Radke-Yarrow, M. (eds), *Development of Antisocial and Prosocial Behavior: Research, Theories and Issues,* New York: Academic Press.

Hinde, R. A. (1987), *Individuals, Relationships and Culture,* Cambridge: Cambridge University Press.

Hinde, R. A. and Bateson, P. (1984c), "Discontinuities versus continuities in behavioural development and the neglect of process," *International Journal of Behavioral Development,* 7, 129–43.

Hinde, R. A. and Barden, L. (1985), "The evolution of the Teddy Bear," *Animal Behaviour,* 33(4), 1371–73.

Hinde, R. A. and Dennis, A. (1986), "Categorizing individuals: an alternative to linear analysis," *International Journal of Behavioural Development,* 9 (1), 105–20.

Hinde, R. A. and Spencer-Booth, Y. (1968), "The behaviour of group companions towards rhesus monkey infants," *Animal Behaviour,* 16, 541–57.

Hinde, R. A. and Stevenson-Hinde, J. (1987), "Interpersonal relationships and child development," *Developmental Review,* 7 (1), 1–21.

Hinde, R. A. and Tamplin, A. (1983), "Relationships between mother-child interaction and behaviour in preschool," *British Journal of Developmental Psychology,* 1, 231–57.

Hinde, R. A., Stevenson-Hinde, J. and Tamplin, A. (1985), "Characteristics of 3- to 4-year-olds assessed at home and their interactions in preschool," *Developmental Psychology,* 21, (1), 130–40.

Hinde, R. A., Titmus, G., Easton, D. and Tamplin, A. (1985), "Incidence of 'friendship' and behavior toward strong associates versus nonassociates in preschoolers," *Child Development,* 56, 234–45.

Jacklin, C. N., Di Pietro, J. A. and Maccoby, E. E. (1984), "Sex-typing behaviour and sex-typing pressure in child/parent interaction," *Arch. Sexual Behaviour,* 13, 413–25.

Jaspers, J. M. F. and Leeuw, J. A. de (1980), "Genetic-environment covariation in human behaviour genetics," in van der Kamp, L. J. T. *et al.* (eds), *Psychometrics for Educational Debate,* Chicester: Wiley.

Kagan, J. (1978), "Continuity and stage in human development," in Bateson, P. P. G. and Klopfer, P. H. (eds), *Perspectives in Ethology (vol. 3), Social Behavior,* New York: Plenum.

Kagan, J. (1980), "Four questions in psychological development,"

International Journal of Behavioral Development, 3, 231–41.

Kroodsma, D. E. and Miller, E. H. (1982), *Acoustic Communication in Birds, (vol. 2)*, New York: Academic Press.

Lazar, I. and Darlington, R. B. (1982), "Lasting effects of early education," *Monographs of the Society for Research in Child Development*, 47, Serial No. 195.

Langlois, J. H. and Downs, A. C. (1980), "Mothers, fathers and peers as socialization agents of sex-typed play behaviour in young children," *Child Development*, 51, 1237–47.

Lehrman, D. S. (1953), "A critique of Konrad Lorenz's theory of instinctive behaviour," *Quart. Rev. Biol.* 28, 337–63.

Lehrman, D. S. (1970), "Semantic and conceptual issues in the nature-nurture problem," in Aronson, L. R., Tobach, E., Lehrman, D. S. and Rosenblatt, J. (eds), *Development and Evolution of Behaviour*, San Francisco: Freeman.

Lenneberg, E. H. (1967), *Biological Foundations of Language*, New York: Wiley.

Lorenz, K. (1950), "Ganzheit und Teil in der tierischen und menschlichen Gemeinschaft," *Studium Generale*, 3–9.

Lumsden, C. J. and Wilson, E. O. (1981), *Genes, Mind and Culture*, Cambridge, Mass: Harvard University Press.

Maccoby, E. E. and Martin, J. A. (1983), "Socialization in the context of the family: parent-child interaction," in Mussen, P. (ed.), *Handbook of Child Psychology, (vol. IV)*, pp. 1–102, New York: Wiley.

Main, M. and Stadtman, J. (1981), "Infant response to rejection of physical contact by the mother: aggression, avoidance and conflict," *Journal of the American Academy of Child Psychiatry*, 20, 292–307.

Main, M. and Weston, D. (1981), "The quality of the toddler's relationship to mother and to father: related to conflict behavior and the readiness to establish new relationships," *Child Development*, 52, 932–40.

Main, M. and Weston, D. R. (1982), "Avoidance of the attachment figure in infancy: descriptions and interpretations," in Parkes, C. M. and Stevenson-Hinde, J. (eds), *The Place of Attachment in Human Behavior*, New York: Basic Books.

Manning, A. (1967), "Pre-imaginal conditioning in Drosophila," *Nature*, 216, 338–40.

Marks, I. M. (1969), *Fears and Phobias*, New York: Academic Press.

Meyer-Bahlburg, H. F. L. (1984), "Gender development: social influences and prenatal hormonal effects," Introduction. *Arch. Sexual Behaviour*, 13, 391–3, and other papers in this issue.

Money, J. and Ehrhardt, A. A. (1972), *Man and Woman, Boy and Girl*, Baltimore: John Hopkins University Press.

Mussen, P., Eichorn, D. H., Honzik, M. P., Bieber, S. L. and Meredith, W.

M. (1980), "Continuity and change in women's characteristics over four decades," *International Journal of Behavioral Development*, 3, 333–48.

Perret-Clermont, A-N. (1985), *Psychological processes, operatory level, and the acquisition of knowledge*, Interactions Didactiques 2 Bis, Universities of Geneva and Neuchatel.

Perret-Clermont, A-N. and Brossard, A. (1985), "On the interdigitation of social and cognitive processes," in Hinde, R. A., Perret-Clermont, A-N. and Stevenson-Hinde, J. (eds), *Social Relationships and Cognitive Development*, pp. 309–27. Oxford: Clarendon Press.

Perret-Clermont, A-N., Brun, J., Saada, E. H. and Schubauer-Leoni, M. L. (1984), "Learning: a social actualization and reconstruction of knowledge," in Tajfel, H. (ed.), *The Social Dimensions (vol. 1)*, Cambridge: Cambridge University Press and Maison des Sciences de l'Homme.

Plomin, R. and DeFries, J. C. (1983), "The Colorado adoptive project," *Child Development*, 54, 276–89.

Quinton, D. and Rutter, M. (1986), *Parenting Breakdown: Making and Breaking Intergenerational Cycles*, Aldershot, Hants: Gower.

Radke-Yarrow, M. and Sherman, T. (1985), "Interaction of cognition and emotions in development," in Hinde, R. A., Perret-Clermont, A-N. and Stevenson-Hinde, J. (eds), *Social Relationships and Cognitive Development*, 173–90. Oxford: Clarendon Press.

Richman, N., Stevenson, J. and Graham, P. J. (1982), *Preschool to School: a behavioral study*, London: Academic Press.

Rutter, M. D. (1987), "Continuities and discontinuities from infancy," in Osofsky, J. (ed.), *Handbook of Infant Development*, pp. 1256–96, New York: Wiley.

Rutter, M. D., Quinton, D. and Liddle, C. (1983), "Parenting in two generations: looking backwards and looking forwards," in Madge, N. (ed.), *Families at Risk*, London: Heinemann.

Scarr, S. and Kidd, K. K. (1983), "Developmental behavior genetics," in Mussen, P. (ed.), *Handbook of Child Psychology (vol. II)*, pp. 345–434, New York: Wiley.

Scarr, S. and McCartney, K. (1983), "How people make their own environments: a theory of genotype-environment effects," *Child Development*, 54, 424–35.

Simpson, A. and Stevenson-Hinde, J. (1985), "Temperamental characteristics of three- to four-year-old boys and girls and child-family interactions," *Journal of Child Psychology and Psychiatry*, 26, (1), 43–53.

Sroufe, L. A. (1979), "The coherence of individual development," *American Psychologist*, 34, 834–41.

Sroufe, L. A., Jacobvitz, D., Mangelsdorf, S., DeAngelo, E. and Ward, M. J.

(1985), "Generational boundary dissolution between mothers and their preschool children: a relationship systems approach," *Child Development,* 56, 317–25.

Stevenson-Hinde, J. (1985), *The development of individual characteristics, with special reference to shyness,* Paper presented at the Workshop on Temperament and Development in Childhood. Leiden, The Netherlands.

Tajfel, H. (1978), *Differentiation between Social Groups,* London: Academic Press.

Tauber, M. A. (1979), "Sex differences in parent-child interaction styles during a free-play session," *Child Development,* 50, 981–8.

Thomas, A. and Chess, S. (1982), "Temperament and follow-up to adulthood," in *Temperamental Differences in Infants and Young Children,* Ciba Foundation Symposium 1989.

Thorpe, W. H. (1961), *Bird-song: The Biology of Vocal Communication and Expression in Birds,* Cambridge: Cambridge University Press.

Thorpe, W. H. and Jones, F. G. W. (1937), "Olfactory conditioning and its relation to the problem of host selection," *Proceedings of the Royal Society,* B 124, 56–81.

Tinbergen, N. (1963), "On the aims and methods of ethology," *Zeits Tierpsychologie,* 20, 410–33.

Trivers, R. L. (1974), "Parent-offspring conflict," *Amer. Zool.* 14, 249–64.

van Hooff, J. A. R. A. M. (1972), "A comparative approach to the phylogeny of laughter and smiling," in Hinde, R. A. (ed.), *Non-Verbal Communication,* pp. 209–42. Cambridge: Cambridge University Press.

Vygotsky, L. S. (1978), in Cole, M., John-Steiner, V., Scribner, S. and Souberman, E. (eds), *Mind in Society: The Development of Higher Psychological Processes,* Cambridge, Mass: Harvard University Press.

Williams, J. and Giles, H. (1978), "The changing status of women in society: an intergroup perspective," in Tajfel, H. (ed.), *Differentiation between social groups,* pp. 431–46. London: Academic Press.

Williams, J. E. and Best, D. L. (1982), *Measuring Sex Stereotypes,* Beverley Hills: Sage.

9 Causes and reasons in social development

Ivana Markova

The term "social development" denotes a broad range of phenomena. For example, social development can embrace processes of child socialization, such as moral development, understanding of the feelings of others, development of intellectual abilities, or the development of the self. Or, by social development one may mean the formation of social representations of commonly shared social phenomena. Thus, it may refer to the changing attitudes to human sexuality resulting from the spread of Acquired Immune Deficiency Syndrome (AIDS); or to changing societal representations of mental handicap resulting from the self-advocacy movement of people with a mental handicap; or to public beliefs about nuclear energy after a catastrophe such as Chernobyl; and so on. At yet another level, social development may occur in a counselling session in which a person suffering from a genetic disease discusses with his or her counsellor the risk of passing that disease on to the next generation. In this social process both participants learn about each other, clarify their positions with respect to the issue in question, cope with feelings aroused by discussing emotionally loaded topics and keep under control conflicts produced by the issues so raised.

These examples do not only differ with respect to whether social development occurs within an individual, dyad, family and societal subgroups. They also differ with respect to the time scales in which such developments take place and the complexity due to the kinds of factors involved. Two questions arise that bear on this tremendous range of phenomena referred to as social development: first, whether it is meaningful to search for principles common to all such phenomena or whether such an exercise would result in theoretically and empirically useless generalities; second, if one subsumes under social development all the above examples, what kinds of phenomena are *not* social development? Let us start with the latter question, the answer to which, it is to be hoped, will also help us to answer the former question. I shall distinguish *human social development* from other phenomena on the basis of three principles.

First, for development of an organism to take place there must be mutual interdependence between that organism and its environment. Thus, every instance of development is *co-development* of organism and environment, and if "development" occurs in only one counterpart it cannot be called development. For example, the sheer unfolding of an organism according to a program should such a thing ever occur, would not be called development. Some cognitive psychologists, e.g. Chomsky (1980) or Pylyshyn (1980), of course, do conceive of development as the unfolding of a genetic program. This point of view assumes that the role of the environment is to trigger off a program which then runs out of its own resources. No assumptions are made either about the changes in the environment due to the developmental changes in the organism, or about the changes in the organism due to the effect of the environment. The principle of mutual interdependence between organism and environment, as we shall see in this chapter, rejects this point of view as ignoring the very essence of evolution and, therefore, as being reductionist.

Second, development involves progressive differentiation of an organism's structures and processes, from global and less structured to more structured and hierarchically organized. In the process of development some structures and processes become relatively stabilized, while others remain relatively unstable, and thus the source of change. The relationship between stable and unstable phenomena give rise to such oppositions as tradition and novelty, the inherited and the acquired, or the old and the new. According to the principle of progressive differentiation, the most stable structures and processes are at the bottom of the organizational hierarchy and are least susceptible to change, while variable structures and processes are at the top of the hierarchy. This principle, therefore, excludes those happenings that are unstructured and random, with no tradition established, with no relationships formed among the phenomena in question. For example, a schizophrenic conversation, consisting of utterly arbitrary utterances, with no meaningful content and no message given to the other participant in conversation, would be nondevelopmental. This claim, however, does not preclude the very same conversation from being conceived as part of a developmental process at another level of analysis, for example, as part of the patient's therapy. Development is always multi-level, and the boundaries within which any discussion proceeds must be properly delineated. Thus, in the case of the above schizophrenic conversation, while the analysis of the sequence of turns between the patient and therapist might not reveal any developmental progress, the conversation as a whole might be conceived as a developmental step in the therapeutic process at another level of analysis. In other words, while an event in isolation might, for some purposes, be conceived as not a development, in another context it could be a stage in development and its developmental character might become apparent when different factors are taken into consideration.

Under certain circumstances the interdependence between individual and organism may break down and regression to a previous developmental stage take place. For example, an organism may be unable to adapt to the changing environment, or an organism may attempt to introduce change ignoring the environmental rigidity. However, if under such circumstances regression occurs, it is never just a case of falling back to an earlier stage. Just as time does not go back, so regression is never a copy of a stage that an organism was in before. Conceptualized within the framework of the principle of progressive differentiation based on stability and variability, regression is always in some ways different from its counterpart in the organism's earlier development or history. From this point of view, even a regression is development of a certain kind.

These two principles, namely mutual interdependence between organism and environment and the principle of progressive differentiation based on stability and variability in phenomena, are essential features of development in general. Human social development, in addition to these two princip.es, is characterized by the relationship between consciously monitored actions and unconsciously occuring events. Self- and other-awareness are reflexive phenomena enabling organisms to evaluate their own mental processes and actions, and those of others. Self- and other-aware individuals monitor their activities, carry out intentional actions and plan for the future. Self- and other-awareness emerge in the process of biological evolution and they have been well documented in non-human primates. However, it is the extent and multitude of the varieties in which reflexive self- and other-awareness in humans are manifest that make it a major characteristic of human social development. The emergence of speech, written language, conceptual thought, judgment and intentional action is intrinsically bound to reflexive self- and other-awareness in the history of mankind and in child development. Yet, paradoxically, as these processes become part of the individual's makeup, of culture, tradition and everyday social reality, individuals use them largely unreflectively, often unaware of the meaning of their speech and actions, and of their effect upon others. The relationship between consciousness and unconsciousness, therefore, will be discussed here as a third principle of human social development. In this view, developmental phenomena that proceed exclusively *below*, and cannot be brought *to* the level of consciousness, will not be called human social development. Examples of such processes would be behavior under hypnosis or unconscious psychological defences.

In sum, human social development, which we shall discuss in this chapter, is defined on the basis of three principles, all of which involve interaction of mutually interdependent dichotomies: *organism and environment, stability and variability,* and *consciousness and unconsciousness.* Phenomena that do not involve the functioning of these three dichotomies will not be called human social development. This definition thus answers our second question, posed at the beginning of this chapter, namely, the question as to what kind of phenomena are *not* forms of social development. Having answered this question, I would like to turn to the first question, as to whether, in characterizing human behavior in terms of very general principles, one's effort results in theoretically and empirically useless generalities. It is our purpose in this chapter to show that it is worthwhile to define human social development in terms of the three principles outlined above.

Organism and environment

Development in general, and human social development in particular, is based on mutual interdependence between organisms and their environment. Natural organisms such as plants and animals, and social organizations such as language and ethical systems, always emerge together with what environs them. What existed before the organism or organization in question emerged was only a potential environment (Gibson, 1979) but not the environment of the organism or organization in question (for more on this issue, see Costall, 1986). An organism and its environment come into existence together, just like a father and his son. Until his son was born, this man was not a father. Nevertheless, he still had a variety of relationships with other people. He might have been a brother to another man, a husband to a woman, a boss to his employees, and so on (Hegel, 1812). "Organism–environment" and "father–son" are relational terms and one component in these dyads cannot be properly understood in isolation from the other component. The intelligibility of the term "organism" in the dyad "organism–environment" is dependent on the intelligibility of its counterpart, "environment". If one component of the dyad develops, it is logically necessary that the other component co-develops.

Many theories, however, have not fully grasped the idea of development as co-development of organism and environment. For example, biological development has been viewed as a one-sided adaptation of the organism to its environment. The environment has often been conceived as a relatively stable entity posing problems for the organism. According to this view, in order to survive the organism must find ways of resolving such problems, either by trial and error or by rational problem-solving activities (for criticism of such views see Lewontin, 1982; Goodwin, this volume; Markova, 1987a). The tendency to view social change as a one-sided influence of powerful majorities has marked post-war American social psychology. As Moscovici (1976) pointed out, theories of social influence in the USA after the Second World War were based almost exclusively on the study of influence as an asymmetric process. It was assumed that the influential majority, because it has power and access to information, affects the minority, which yields to the pressure of that power and its information. In other words, just as biological development has been viewed in terms of the adaptation of an organism to its stable environment, so has social change been viewed in terms of minorities conforming to powerful majorities. Just as in the theory of biological development the organism's environment has been conceived as stable, so in the theory of social change, the minority's environment, i.e. the majority, has been conceived as stable, leaving the minority powerless.

Another example of a one-sided theory in human social development is *individualism*. Individualism has been deeply rooted in European culture since the Renaissance. It has attempted to explain the individual's intellectual, emotional and social development in terms of Cartesian innate dispositions, Lockean simple and complex ideas of sensation, and Kant's a priori modes of knowledge, all of them ignoring the individual's social environment. In the research that has been based on this tradition such essential aspects of human social development as speech, concept formation and moral development, have been conceived in terms of biological unfolding of a genetic program or in terms of biological maturation. Much of the past and present cognitive psychology belongs to this tradition of thought.

In sum, one-sided theories of development in general, and of human social development in particular, have penetrated mainstream psychological thinking since the beginning of psychology as a biological and social science. At the same time, critics of such theories called for a reconceptualization of development in terms of mutual interdependence between organism and environment.

At the turn of this century, James Mark Baldwin strongly objected to a one-sided view of human social development. For him, social progress was based on a "dialectic of social growth", a give-and-take relationship between the individual and society. He argued that the form of collective organization cannot be social without having first been individual. At the same time, the matter of social organization cannot be individual without having first been social (Baldwin, 1897, p. 570). Baldwin maintained that social progress at any level of development involves interaction between social and individual, that is, between the general and particular, which mutually determine each other. Pure collectivism deprives theories of human social development of the original ideas and creative contributions of individual minds. Pure individualism, on the other hand, dissolves the achievement of social history and leaves the person a human atom (Baldwin, 1911). According to individualistic theories of knowledge, a person is assumed to start in isolation with his or her sensations and cognitions, and then come to some sort of agreement by "matching" his or her world with those of other individuals. Such theories that ignore the social origin of knowledge, "have to be laid away in the attics where old intellectual furniture is stored" (Baldwin, 1910, p. 78). According to Baldwin human knowers start with shared experiences of the world, and on the basis of these they develop their own independent thoughts. Unfortunately, it was his own social theory of knowledge that was put away in the attic so that it might leave room for the growing positivism and individualistic epistemology in the early years of this century. As a result, it is only very recently that Baldwin's work has been rediscovered (Russell, 1978; Broughton and Freeman-Moir, 1982).

George Herbert Mead (1934), some years later, expressed similar ideas concerning the relationship between organism and environment, and on the social nature of knowledge and communication. In biological development, he argued, organism and environment are mutually dependent, and determine each other. The individual organism "selects that to which it responds" and then uses it for its own purposes (Mead, 1934, pp. 245–6). Similarly, in human social development one must start with the individual–society dyad. Mead was very critical of what he called the philologist's approach to the study of language:

The philologist ... has often taken the view of the prisoner in a cell. The prisoner knows that others are in a like position and he wants to get in communication with them. So he sets about some method of communication, some arbitrary affair, perhaps, such as tapping on the wall. Now each of us, on this view, is shut up in his own cell of consciousness, and knowing that there are other people so shut up, develops ways to set up communication with them. (Mead, 1934, p. 6)

According to the philologist approach, the individual consciousness is prior to what is publicly shared. Instead, Mead argued, language, knowledge, and logical thought develop through internalization of socially shared experiences. Society and individual comprise a dyad that is prior to both the individual and society taken as isolated existences.

Mutual interdependence between organism and environment as a principle of development is also assumed by Moscovici (1976) in his interactionist model of social change. In Moscovici's theory, majority and minority form a dyad mutually influencing each other. According to this theory, every individual is acted upon by others and likewise acts upon them. One cannot separate the emission from the reception of influence, nor should we fragment these two aspects of a single process by attributing the one exclusively to one partner (majority), and the other to the other partner (minority) of social interaction (Moscovici, 1976, p. 68).

The mutuality of dependence between individual and environment shows itself remarkably well in the study of the history of ideas. Scientific discoveries, artistic creations, new ideologies and changes in traditions are possible only if both the individual and societal factors contribute to that process. In his study of the structure of scientific revolutions, Kuhn (1962) has shown that preconceived systems of ideas, or paradigms, are so influential that they determine the kinds of theories proposed by scientists. Thus he gives a variety of examples to that effect. For instance, although all knowledge necessary for the discovery of the pendulum was available in the Aristotelian era, it was only in the sixteenth century, two thousands years later, that the pendulum was discovered by Galileo when he focused on different kinds of concepts than did the Aristotelians. Or, although since 1690 astronomers had seen stars in the positions of Uranus, they did not identify them with a planet because their minds did not operate in a framework allowing the existence of another planet. It took nearly a hundred years for Uranus to be discovered. In a similar vein, Nicolson (1950) discusses the powerful metaphor of the Circle of Perfection and its effect on Western culture from the Ancient to the Modern Age. Western civilization, she pointed out, had been based on the idea that God created all things in the universe, including the human body, as circular as possible. Just as the celestial bodies were believed to move in perfect circles so the motion of the human soul was believed to be circular. The idea that a human being and his or her soul is a microcosm mirroring and uniting in him- or her-self the terrestrial, celestial and spiritual worlds was quite common in medieval and Renaissance culture. The circle was a symbol of God and implied both the beginning and the end of the world. The idea of the Circle of Perfection penetrated not only scientific views such as the motion of heavenly bodies, but also the literature, art and poetry. Nicolson has drawn attention to numerous images of the circle in poetry, celebrating its perfection and beauty of form, its infiniteness and eternity. So powerful was the image of the Circle of Perfection that Kepler who broke it and substituted elliptical movement of celestial bodies for circular, did it only with great reluctance. Nicolson paraphrases Kepler's feelings as follows:

Circular motion still remains the perfect motion, and the circle is always a symbol of God. If the planets do not move in circles, the limitation is not in the Creator but in the *creature*; the planets sought the circle, but in so far as they are not only spirit but matter, possessing limitations of grossness not shared by the Creator who is pure spirit, they move not in perfect circles but in ellipses, "imitating" so far as their natures permit "the beauty and the nobleness of the curved." (Nicolson, 1950, p. 134)

One can thus see that the breaking the circle was not just a substitution of one scientific theory by another. It was also the breaking of a culture, of a tradition that was shared by the whole community, and this community strongly resisted the changes that appeared to threaten its very existence.

In contrast to examples showing the effect of the social and cultural environment on the kinds of ideas produced at certain times, other examples show individuals producing ideas reaching far beyond those acceptable or comprehensible in the social environment in which that individual lives. There are numerous cases of creative individuals in science who were totally ignored or rejected or were even persecuted for producing such ideas. Gregor Mendel's discovery of the statistical laws of inheritance were totally ignored when presented in 1865 to the natural history society in Brno and then published one year later. Koestler (1964) refers to martyrs of science whose lives ended tragically. For example, the co-discoverer of the principle of the conservation of energy, Robert Mayer, and Ignaz Semmelweiss, who dramatically reduced childbed fever by introducing the strict rule of washing hands in chlorinated lime water, both ended insane.

Evidence showing the necessity of mutual interdependence between organism and environment for social change to occur is provided by Gruber (1974) in his analysis of the development of evolutionary theory by Darwin. Gruber views a creative process as co-development of the individual's scientific ideas and of the social environment in which the individual lives. Creative thinking is not the isolated act that it is often held out to be by laypersons and by students of creativity. It is a developmental process in which creative individuals must cope with various kinds of social pressure, such as the threat of persecution, the pressure of existing norms, and the difficulties of escaping from accepted ways of thinking and formulating problems. Gruber's analysis shows how Darwin himself carefully monitored what to say and when to say it because of fear of persecution and his own uncertainties regarding his ideas.

We find a similar relationship between society and the individual in the world of art. The interdependence between the artist's painting style and the established pictorial representations of the world was explored by Gombrich (1960). History of art as a subject of study is possible only because there are, in different societies and cultures, particular representations of the world that are reflected in the artists of that period. We easily recognize a Chinese from a Dutch landscape, just as we do portraits painted in different historical periods. The style within which an artist operates limits his or her choice and restricts the possibilities for particular representations of the world. In fact, if everything were possible without limitation in every artistic period communication would break down: "It is because art operates with a structured style governed by technique and the schemata of tradition that representation could become the instrument not only of information but also of expression" (1960, pp. 319–20). Art that lies outside the latitude of acceptance of the social representation of a particular historical period is ignored or rejected. For example, in the Enlightenment, while the art of ancient Greece was evaluated as the height of harmony and beauty, medieval culture was evaluated only negatively. Gothic culture was viewed as uncivilized by the Enlightenment and it was only with romanticism that medieval art was re-evaluated, and the term "Gothic architecture" given the meaning it has today (Mead, 1936). But even the "visual discoveries" (Gombrich, 1982), of the Impressionists, which focused on colored reflections and shadows, were not accepted by the public when they first appeared, and the public had to learn to recognize them as familiar. These examples show that the personal vision of an artist and tradition are mutually inter-dependent. The history of social representations in art is also a history of the choices available to individual artists. It is within these limits that they co-develop.

Stability and variability

The process of development is characterized by a pair of tendencies functioning in opposite directions from each other: first, a tendency to remain in the existing state, and second, a tendency to change. Many developmental models assume that a tendency towards stability and a tendency towards change are hierarchically organized. Thus, Simon (1962), who developed an architectural model of complexity argues that the most efficient complex systems are based on hierarchical arrangements of stable and variable phenomena. In developing systems, Simon argues, a configuration of stable building blocks provides the raw material from which new structures can develop by trial and error. While Simon uses biological evolution as a test of his model, he applies it to all kinds of developing complex systems, including social institutions, economic establishments, human problem-solving and the acquisition of knowledge, and in systems of artificial intelligence. Similarly, Bronowski's (1970) model of stratified stability states that, at every level of evolution, relatively stable units, e.g. atoms, come together by chance to form more complex units, such as molecules. If molecules are sufficiently stable then they, in turn, become the building blocks for more complex configurations, and so on. In organic evolution, according to Bronowski, the same principle applies. Genes are relatively stable structures and building blocks for more complex organic configurations. Cells are stable structures at a higher level and so are organs at yet a higher developmental level. Jantsch (1979) expresses a very similar idea of development. In sum, relatively stable structures represent a tendency towards stability. They are building blocks for new structures. The tendency towards stability is the tendency of the organism or system to preserve its identity. The other tendency leads the organism or system towards change. This tendency functions in the opposite direction from stability, driving the organism or system to express its individuality, volition, desires and spontaneity.

Although the principle of mutual relationship between stability and variability has been assumed in developmental models based on system theories, in psychological theories the developmental tendencies towards stability and towards change have often been treated in isolation from each other. For example, certain theories of the development of intelligence focus attention on heredity, that is, on relatively stable and innate factors in the development. Other theories over-emphasize the function of variable circumstances in the development of intelligence and underplay the role of stable factors. In a similar vein, some theories of personality focus attention on stable traits while others focus on situations; and so on. Such separation of the two components of the dyad stability–variability in psychological theories is due to difficulty in the conceptualization of phenomena in relational terms (Markova, 1987a).

However, although much mainstream psychology has treated stability and variability in developmental theories in separation from each other, there have been exceptions. Thus, Baldwin (1897) defined biological and social development in terms of two opposite and complementary tendencies, *habit* and *accommodation*. *Habit* is the organism's conservative tendency to remain in the state in which it is already. It is a tendency to recapitulate more and more readily processes and structures that are vitally beneficial for the functioning of the organism. If the organism encounters new phenomena in the environment it tends to cope with them in terms of its existing structures and processes. Indeed, new phenomena make sense for the organism only to the extent to which the organism can cope with them in terms of its existing structures and processes. For example, an event is perceived as threatening for the organism only to the extent to which it can be anchored to something that already presents a threat for that organism. The other tendency of the organism, quite opposite to habit, is *accommodation*. Accommodation is the tendency of the organism to cope with new phenomena by performing more complex functions (Baldwin, 1894, p. 479). It is the organism's openness towards change, trying out new activities, learning, and changing its structure and processes. All capacities that the organism learns are examples of accommodation. Baldwin points out that accommodation is opposed to habit in two ways. First, accommodation is directed towards the organism's future, or as Baldwin (1894, p. 478) puts it, it has a prospective reference.

Habit, on the other hand, has a retrospective reference, relying upon past and old movements of the organism. Thus, accommodation always runs ahead of habit. Second, accommodation, because it involves the selection of new activities, tends to get into direct conflict with old habitual activities, and thus breaks habits.

Habit and accommodation are relational phenomena. They are interdependent and their interaction "gives rise to a two-fold factor in every organic activity of whatever kind" (Baldwin, 1894, p. 481). Habit is constantly modified by accommodation and accommodation is restricted by habit. Each function of the organism, Baldwin argues, can be understood only in terms of this two-fold factor, habit–accommodation, whether it is attention, instinct or emotion. Consider the interdependence of habit and accommodation in Baldwin's conceptions of the historical development of ideas. Thus he pointed out that society has its own *social habit*, its conservative spirit guarding it against agitators, innovators and advocates of change. Its established traditions and institutions are considered to be safe because they have already been tested and experienced. But society conserves and guards not just habits, traditions and institutions, but also the conservative attitude of mind, that is, what is known as "public opinion" (Baldwin, 1897, p. 184). Yet, if only habit were in operation, there would be no social progress. Just as in biological development an organism accommodates to new and changing conditions of the environment "sometimes indeed working directly in opposition to the habits already acquired", so also it is with the social body (p. 185). Both in the development of an individual and of society social accommodation operates and secures change. Any innovation, however, is always rooted in the already acquired habits and knowledge. It is impossible, Baldwin argues, to make an invention that would totally break with the past culture and tradition. There is always a continuity of knowledge and even such revolutions as Copernican theory drew upon the data of common knowledge. Baldwin's theory of habit and accommodation is thus in full agreement with those of Kuhn, Gombrich, Moscovici and Gruber discussed in the previous section of this chapter.

Baldwin's idea of habit and accommodation as a principle of development was taken over by Piaget (1953) in order to explain the development of intelligence. In this effort, however, Piaget restricted himself to the study of the development of logical and mathematical thought in the child. Baldwin, however, attempted to study development as such, and not just some of its aspects. He took, as Freeman-Moir (1982) pointed out, a "decidedly Hegelian sweep", trying to explain the nature of mind and to suggest a theory of development in the Hegelian manner. This sweep, as often happens to grand theories, did not come off, and Baldwin's work, as we have already remarked was forgotten until very recently.

The principle of stability and variability in Baldwin's and Mead's theories of self-consciousness

James Mark Baldwin (1861–1934) and George Herbert Mead (1863–1931) were contemporaries and they postulated very similar theories of the development of the self and of self-consciousness. The similarity of their theories is probably due to the fact that they were both strongly influenced by Hegel's theory of the development of self-consciousness (on Hegel and Mead, see Markova, 1982; also Markova, 1987a). Yet, Baldwin and Mead rarely referred to each other's work. The reason for this mutual neglect could be partly due to the fact that Baldwin's main works were published at the turn of the century. He was forced to resign from academic life in 1908 because of "misbehavior" and from 1912 until his death he lived in France. After he left the United States his scientific contributions were quickly forgotten. Mead's papers concerning his theory of the self and of self-consciousness first appeared in the press not long before Baldwin disappeared from academic life. In *Mind, Self and Society* (Mead, 1934) there is only one insignificant reference to Baldwin.

In Baldwin's conception, the process of formation of the self is based on the "dialectic of personal growth". By dialectic of personal growth Baldwin means a process of mutual interdependence between the *self* and *other selves*. The individual's awareness of the self (*ego*) and of the other (*alter*) arise together. As Baldwin puts it, both the *ego* and *alter* are originally crude and unreflective and largely organic. However, they get "purified and clarified" in the process of mutual interaction: "my sense of myself grows by *imitation* of you, and my sense of yourself grows in terms of my sense of myself" (Baldwin, 1897, p. 15; my emphasis). Thus, ego and alter are essentially social. They grow from organic and unreflexive to social and reflexive phenomena through a give-and-take relationship which manifests itself as *imitation*. Baldwin distinguishes three stages in the process of self- and other-consciousness in child development. First, the child learns to distinguish between persons and things, and between different persons and different things. For example, the child recognizes whether he or she is held by the mother or by the nurse. The second stage of the development of the self is characterized by the child's acquisition of a sense of personal agency. The child becomes aware of his or her own accommodations, he or she is able to carry out intentional actions, to accomplish his or her desires and to make distinctions between the activities of different persons. Finally, the child becomes aware of others as *me's*, that other people have their selfhood.

The growth of ego and alter is based on *habit* and *accommodation*. Just as with other kinds of human social development, the development of the self involves a *conflict*, in this case of the self with other selves: "The self meets self, so to speak" (Baldwin, 1894, p. 342). In this process, the self of accommodation, that is the self that learns and changes, collides with the self of habit, of character and of desire to dominate others. In Baldwin's words, *the self of personal agency* gets into conflict with *the social self*. One can thus see that the relatively *stable* component of the development of the self is represented by inborn impulses, automatized responses and natural desires. Accommodation, the relatively *variable* component of the self, is represented by the social self, by the individual's ability and tendency to sympathise with others and to take their perspective. These two components of the self, personal agency and the social self, are intrinsically related. As the self develops through *imitation*, new structures and processes of the social self are incorporated into the older ones of personal agency. In this process accommodation turns into habit, or, as Baldwin puts it, "accommodation, *by the very reaction which accommodates*, hands over its gains immediately to the rule of habit" (Baldwin, 1894, p. 480).

Mead's (1934) theory of the self, rather than being based on *imitation*, is based on *communication*, or, in Mead's words, on the interaction of gestures between participants. In this conception, the individuals' awareness of themselves and of others develop concurrently in the process of communication through mutual adjustment of these individuals to each other. The individuals' gestures, in their phylogenetic and ontogenetic origin, are unreflective, i.e. *non-significant*. There are originally no common or shared meanings of these gestures. For example, a child might cry because he or she is in pain. The mother comes and soothes the pain. The child at this stage, however, does not cry *in order* to express his or her discomfort or *in order* to attract the mother's attention. The child's response to pain is natural and instinctive, and to explain this response one does not need to refer to self- or other-awareness. It is only when the child *intends* to attract his or her mother's attention and has certain *expectations* about the kind of response from the mother that one needs to think in terms of reflexive self- and other-awareness. It is only when one can assume shared meanings between participants that gestures are reflexive, and therefore *significant*. Thus, in Mead's conception, just as in that of Baldwin, one's awareness of the self emerges together with one's awareness of the other.

The development of the self in Mead's theory is based on the interaction of the two components of the self, the *I* and the *Me*. The *I* is a spontaneous, unpredictable and innovative component of the self. It is the *agent* and *experiencer*: "That action of the 'I' is something the nature of which we cannot tell in advance" (Mead, 1934, p. 177). The *I* gives the self a sense of freedom and initiative and a possibility to respond in novel ways to social situations. In contrast, the *Me* is a habitual and conventional component of the self. It arises through the progressive ability of the individual to take the attitude of others towards him- or her-self, that is, to perceive him- or her-self as an *object*, in the way others do. It is when the individual becomes aware of him- or her-self as an object that he or she becomes truly self-conscious. In this sense, the Me is the means of social regulation and self-monitoring. It sets limits on the activities of the *I*. The *Me* in its most generalized sense represents society's conventions, opinions, morals and sets of organized attitudes. Social control operates to the extent to which the individual assumes the attitudes of those in his or her group and with whom he or she is involved in social activities (Mead, 1924/1964, p. 290).

Just as Baldwin's *habitual personal agency* and *accommodative social self* are relational phenomena, so are Mead's *I as agent*, and *Me as social object*. Mead's I as an agent is practically involved in the world and it changes into the Me once the action is carried through or the word expressed so that it can be reflected upon. The I cannot appear in the self's consciousness *as* I but *only* as Me (Mead, 1913/1964). The I exists only in memory:

We can go back directly a few moments in our experience, and then we are dependent upon memory for the rest. So that the 'I' in memory is the spokesman of the self of the second, or minute, or day ago. As given, it is a "me" which was the "I" at the earlier time. (Mead, 1934, p. 174)

Thus the I, the agent component of the self, becomes part of old, habitual structures. The old structures then take on new characteristics, adjusting and readjusting to their ever-changing social environment. The self, in this sense, does not reach a state of equilibrium, but the I is always ahead, living in the future.

In sum, Baldwin's and Mead's theories of the development of the self are similar in the following respects. First, they both assume that self- and other-awareness emerge together from unreflective social behavior. Second both theories assume that the self develops in the process of interaction between the *personal* and *social* components of the self. Third, both theories assume that the self develops through the interaction of a relatively stable, habitual component and of a relatively variable, accommodative component of the self. They differ, however, in one important respect which raises an interesting theoretical problem.

In Baldwin's theory, *personal agency* is associated with habit, the organic and instinctive nature of the self, and therefore with the *stable* and *conservative* nature of the self. In Mead's theory, in contrast, *personal agency*, the I, is associated with spontaneity, creativity and unpredictability, and therefore with the *variable* and *novel* nature of the self. In Baldwin's theory, it is the *social self* that is associated with invention, creativity and adaptation, i.e. with the *variable* and *novel* nature of the self, whilst in Mead's theory, the *social self*, the *Me*, is associated with habit and convention, i.e. with the *stable* and *conservative* nature of the self. In other words, the two authors view the functions of personal agency and the social self in exactly opposite ways. This, of course, raises the problem as to who is right and who is wrong and there appears to be no easy solution because both theories appear to be consistent with observation.

It seems to me that the problem as outlined in the prevous paragraph arises from the fact that although Baldwin and Mead were both dialectical theoreticians, they did not apply dialectic method thoroughly enough to their theories. Just as the principle of stability and variability defines the nature of developing systems as such, so it defines the nature of their sub-systems, as pointed out in our discussion of hierarchical organizations earlier in this section. Therefore, both personal agency and the social self comprise hierarchical organizations involving stable and variable components. Personal agency, the I, has both habitual, stable components, and innovative, variable components. The agent utilizes both automatized and repetitive responses that are embedded in relatively stable structures and processes, *and* innovative and unpredictable activities that constantly change such structures and processes. The social self, the Me, for its part, involves both a social control rigidly directing the self's activities in terms of old structures, and social sensitivity by which the self accommodates and constantly changes these old social structures. It is through such multi-level processes that both personal agency and the social self transform each other. Baldwin and Mead each focused on one feature only, *either* stability *or* variability, rather than on stability and variability as a dyad of the components of the self. It is in this respect that their dialectics of personal and social growth are incomplete.

Consciousness and unconsciousness

The dichotomies *individual and environment*, and *stability and variability*, are essential to development as such, be it the development of plants or human social organizations. In contrast, *reflexive consciousness* appears to be an exclusive characteristic of the biologically most complex species. Anthropological, ethological and psychological studies have traced the origin of reflexive consciousness in animals and in particular in non-human primates. The beginning of self- and other-awareness in animals, of intentional action and perception of intentionality, the recognition of individuality and the origins of empathy, have been shown to exist in various animal species (Markova, 1987b). However, it is the *quality* and *extent* of reflexive consciousness in humans that makes it a unique characteristic of human social development. Reflexive consciousness in human social development includes the following characteristics: the ability to recognize one's own existence and experience, and the existence and experience of others; knowledge of one's own agency and of that of others; the ability to monitor and evaluate events in one's own life, and to make decisions about one's own future on the basis of that knowledge; and the ability to communicate one's awareness of oneself and of others to other fellow human beings.

Students of the evolutionary nature of consciousness have argued that reflexive consciousness has developed because it is highly efficient in terms of adaptation and survival of species (Crook, 1980; Humphrey, 1983). To be aware of the other implies the possibility of responding to the other as an individual and to his or her idiosyncratic characteristics. Responding to the other as an individual requires a kind of communication that is highly flexible. Flexibility and efficiency of communication appear to be particularly important for the co-operation and complex interaction that occurs in mutually interdependent individuals functioning in social groups (MacLean, 1973; Humphrey, 1983).

Throughout the history of human civilization, reflexive consciousness has undergone remarkable development. For example, historical analyses have shown that different conceptions of the self and of the other have evolved, in close relationship to cultural, societal, and personal values, to the level of education and literacy, the structure of the family and social organizations. Thus, particular cultural and economic conditions in Europe in the last four hundred years have given rise to a kind of self-consciousness that focuses on the self and its relationship to others in a way quite unfamiliar in non-western traditions. For example, while in the western conception the conflict between the self and others is considered essential to self-growth, in Japanese Confucianism the idea of harmony in social relationships and avoidance of conflict appears to be the major principle in the development of the individual's self- and other-awareness (DeVos, 1985).

Studies in child social development show that as the child grows older he or she acquires more complex forms of reflexive consciousness. Thus, empirical research has demonstrated the child's developing ability to be aware of and reflect on the feelings, thoughts, intentions and actions of others; to conceptualize his or her own and other selves; to be aware of him- or herself as an agent and to wish to be recognized as such by others (Markova, 1987b).

Human social development, however, is not just the development and practising of reflexive consciousness. We are born into an existing social world, into existing societal ways of seeing and understanding the world, and into accepted conventions and language. Much of the existing social reality we accept unreflectively, not realizing the effect of this commonly shared and accepted social reality upon ourselves. Indeed, the less aware we are of the influence of social reality upon ourselves, the greater its effect (Moscovici, 1984). Take language, for example. A child is born into society and learns its language, both unreflectively and reflectively. Concerning the former, he or she learns the meanings of words and uses them as others do. Goffman (1968) in his analysis of stigma pointed out that often we use words such as "cripple", "bastard" or "moron" as a source of metaphor and imagery without giving a thought to their original meaning. We do not realize the perpetuating stigmatizing effect such words have on others so labelled. More generally, words and speech actions have diagnoses and prognoses built into their meanings. Our own recent research which I shall describe shortly, demonstrates this point clearly (Markova and Cattermole, 1987).

Speech therapy and training in social skills given to people with a mental handicap is commonly based on a didactic "teaching-type" style. The techniques used in speech therapy and in social skills training involve teaching situations in which a person with a deficit is given instruction, demonstration, reinforcement or feedback on his or her performance, and is assigned homework and practising exercises. The deficit model, by its very nature, is based on asymmetrical communication in which a person with a mental handicap is put into a non-reversible role of the one who *does not know* or who *cannot do* something or other, and thus has to be taught by those who know and who can. We decided, in our own research, rather than use a didactic "teaching-type" style in training in social skills, to make it possible for people with a mild mental handicap to be involved in discussion groups in which they would participate as equal discussants alongside people without a handicap. Thus, two discussion groups consisting of people with a mental handicap and two researchers were run for ten weeks. The researchers were not only particularly aware of the importance of engaging in a symmetrical type of interaction with participants with a mental handicap, but were personally convinced that this was the right procedure to be adopted in speech therapy and social skills training. Only in this way can the confidence of people with a mental handicap in their ability to communicate be increased. It is, we believe, lack of such confidence that is, to a great degree, responsible for their lack of social skills. However, the subsequent analysis of the data, using a Sinclair and Coulhart (1975) type of method, revealed surprising results. We found that the researchers highly dominated the discussion. Moreover, whilst they spoke more than group members, this was not the case for all types of communicative actions. Researchers tended to use a lot of elicitations, evaluations and directives, that is, actions that defined initiations, follow-up moves and that required responses. In contrast, group members with a mental handicap mostly used replies and gave information, that is, actions that denote responses and did not require follow-up. These results were interesting, not so much for what we found, but because what we found was a great surprise for the researchers who were quite unaware of what they did and did not do in the discussion.

In his book, *The Silent Language*, Hall (1959) compared non-verbal communication to the tip of the iceberg: while communicating we are aware of only a small portion of what we actually communicate; a much greater part remains under the level of awareness. Both language and our social representations of societal phenomena, such as mental illness, mental handicap, stereotypes and beliefs, are so powerful and resistant to change because we are so little aware of their effect, and use communicative actions and words that perpetuate existing states without being aware that we are doing so.

The relationship between consciousness and unconsciousness in human social development is also apparent in Baldwin's (1894) and Mead's (1934) theories of development which we have already discussed. Let us consider, first, Baldwin's theory of habit and accommodation. Habit is the tendency to automatize reactions and activities, and as a result the individual's consciousness tends "to evaporate from such reactions" (Baldwin, 1894, p. 228). Consequently, habit leads the individual to a loss of overview of his or her situation because it diffuses concentration and undermines attention. Thus, those reactions and processes that are most dominated by habit, that is, those that are most automatized, that are inherited and instinctive, are also those least controlled by consciousness. In contrast, situations in which the individual must cope with new conditions, and in which he or she experiences emotions such as pleasure, envy or fear, in which he or she is aware of pain or other states which require the focusing of attention, self-control and self-monitoring, are least influenced by habit.

Mead discussed the relationship between conscious and unconscious processes in a similar manner. As he pointed out, reflexive consciousness comes into operation when an individual can no longer carry out automatized activities because their train is interrupted. Reflexive consciousness arises when a person is suddenly faced with a conflict and cannot carry out the act unreflectively and automatically. For example, a person finds him- or herself facing a chasm, and the chasm is too wide to jump over. The person now stops and starts considering the various possibilities available to him or her in order to solve the problem: "there is that checking of activity which is essential to reflective consciousness; the necessity for adjustment to the changed situation" (Mead, 1914/1982, p. 45). Thus, reflexive intelligence is the ability to solve problems on the basis of past experience and bearing in mind the possible future consequences. Reflexive intelligence, as Mead says, involves both memory and foresight (Mead, 1934, p. 100).

Discussing the unconscious in this chapter I have not been concerned with that kind of the unconscious that cannot be brought to consciousness because the human agent has no access to it due to its physiological nature or due to repressed psychopathological defences (von Cranach, 1982). Rather, I have been concerned with unconscious behavior and mental processes that are not essentially different from the conscious ones and that can, therefore, be brought to consciousness relatively easily (see also Mead, 1927/1982, p. 114). Such unconscious behavior can be basically of twofold origin: First, mental phenomena can become unconscious by being submerged under the level of consciousness because the individual has automatized the activities in question.

For example, having mastered a musical instrument or the technique of driving a car, one carries out such activities quite automatically, unaware of one's hand-movements. Under such circumstances, only when one makes an error is one suddenly faced with the problem of getting back into gear. Alternatively, mental phenomena may be unconscious because the individual acquired them unthinkingly, having been born into a particular social reality and having taken their existence for granted, just like that of physical objects, be they trees, cars or computers. For example, one is born into a culture in which the use of a term such as "mentally retarded person" is taken for granted. Thus, one acquires the term and uses it automatically, until one is faced with the individual who is "mentally retarded" and that individual expresses a strong objection to the use of that term. It is only now that one starts thinking, bringing the use of the term "mentally retarded person" into one's consciousness. It is when the public in general raises its awareness of the problem that alternatives to "mental retardation", e.g. "learning difficulty", get into usage.

The relationship between the conscious and unconscious is a unique principle of human social development. Many human social phenomena arise, originally, through reflexive consciousness. A paradigmatic example is language. Language, in its origin, is a product of reflexion (Herder, 1771; Humboldt, 1836). A word, both in its spoken and written form, makes it possible for the individual to detach him- or herself from the immediacy of action and perception by pointing to something, making it possible to interpret that thing and one's own understanding of that thing. Yet, at the same time, language as a societal product is a conservative phenomenon that, to a large extent, is used unreflectively. We often speak, as we have already pointed out, quite unaware of what we say and how we affect other people.

Bringing human social phenomena into the consciousness of others by raising their awareness of them, just like forcing something *out* of consciousness, is one of the main means of *power* in democratic systems. Those minority groups concerned to produce social change and to change the status quo are very much aware of this problem, and their strategy is to raise awareness by pointing to the thing and creating a problem. Once the problem is created it requires solution. Similarly, a child interacting with his or her parents may use the technique of distracting attention in order to get his or her way. Birdwhistell (1970) refers to such examples in his own childhood, using a strategy of "talk-talk-talk" with his parents and thus avoiding "discussion of torn pants or of inadequate report cards". In conclusion, the power to monitor consciousness and unconsciousness is a power to monitor human social developement, at its worst and its best.

References

Baldwin, J. M. (1894), *Mental Development in the Child and the Race*, New York: Macmillan.

Baldwin, J. M. (1897), *Social and Ethical Interpretations in Mental Development*, London: Macmillan.

Baldwin, J. M. (1910), *Darwin and the Humanities*, London: Swan Sonnenschein.

Baldwin, J. M. (1911), *The Individual and Society*, London: Rebman.

Birdwhistell, R. L. (1970), *Kinesics and Context*, Philadelphia: University of Pennsylvania Press.

Bronowski, J. (1970), "New concepts in the evolution of complexity: stratified stability and unbounded plans", *Zygon*, 2, 77–96.

Broughton, J. M. and Freeman-Moir, D. (1982), (eds), *The Cognitive Developmental Psychology of James Mark Baldwin*, Norwood: Ablex.

Chomsky, N. (1980), "Rules and representations," *The Behavioural and Brain Sciences*, 3, 1–61.

Costall, A. (1986), "The 'psychologist's fallacy' in ecological realism," *Theorie & Modelli*, 3, 37–46.

Crook, J. H. (1980), *The Evolution of Human Consciousness*, Oxford: Clarendon Press.

DeVos, G. (1985), "Dimensions of the self in Japanese culture," in Marsella, A. J., DeVos, G. and Hsu, F. L. K. (eds), *Culture and Self*, New York and London: Tavistock.

Freeman-Moir, D. (1982), "The origin of intelligence," in Broughton, J. M. and Freeman-Moir, D. (eds), *The Cognitive Developmental Psychology of James Mark Baldwin*, Norwood N.J.: Ablex.

Gibson, J. J. (1979), *The Ecological Approach to Visual Perception*, Boston: Houghton-Mifflin.

Goffman, E. (1968), *Stigma*, Harmondsworth: Penguin.

Gombrich, E. H. (1960), *Art and Illusion*, London: Phaidon.

Gombrich, E. H. (1982), *The Image and the Eye*, Oxford: Phaidon.

Gruber, H. E. (1974), *Darwin on Man: A Psychological Study of Creativity*, together with Darwin's early and unpublished notebooks, transcribed and annotated by P. H. Barrett, London: Wildhood.

Hall, E. T. (1959), *The Silent Language*, Garden City, NY: Doubleday.

Hegel, G. W. F. (1812–6), *Science of Logic*, trs. Johnston, W. H. and Struthers, L. G., Woking: Unwin, 1929.

Herder, J. G. (1771), *On the Origin of Language*, in Herder, J. G., *Samtliche Werke*, Ed. B. Suphon, reprinted 1967, Hildesheim: George Olins.

Humboldt, W. von (1836), *Linguistic Variability and Intellectual Development*, trs. Buck, E. C. and Raven, F. A., Coral Gables: University of Miami Press (1971).

Humphrey, N. (1983), *Consciousness Regained*, Oxford and New York: Oxford University Press.

Jantsch, E. (1979), *Die Selbstorganisation des Universums*, München: Carl Hanser Verlag.

Koestler, A. (1964), *The Act of Creation*, London: Hutchinson.

Kuhn, T. S. (1962), *The Structure of Scientific Revolutions*, Chicago: University of Chicago Press.

Lewontin, R. C. (1982), "Organism and environment," in Plotkin, H. C. (ed.), *Learning, Development, and Culture*, Chichester and New York: Wiley.

MacLean, P. D. (1973), *A Triune Concept of the Brain and Behavior*, Toronto: University of Toronto Press.

Markova, I. (1982), *Paradigms, Thought and Language*, Chicester and New York: Wiley.

Markova, I. (1987a), "On the interaction of opposites in psychological processes," *Journal for the Theory of Social Behaviour*, 17, 279–99.

Markova, I. (1987b), *Human Awareness*, London: Hutchinson Education.

Markova, I., and Cattermole, M. (1987), "Communicative awareness and group interaction in people with a mild mental handicap," Paper read at the IIIrd International Conference on Language and Social Psychology, Bristol.

Mead, G. H. (1913/1964), "The social self," *The Journal of Philosophy, Psychology and Scientific Method*, 1913, 10, 374–80. Reprinted in Reck, A. J. (ed.) (1964), *Selected Writings of George Herbert Mead*, Chicago and London: University of Chicago Press.

Mead, G. H. (1914/1982), 1914 class lectures in social psychology, in Miller, D. L. (1982) (ed.), *The Individual and the Social Self*, Chicago and London: University of Chicago Press.

Mead, G. H. (1924/1964), "The genesis of the self and social control," *International Journal of Ethics*, 35, 1924–5. Reprinted in Reck, A. J. (ed.), *Selected Writings of George Herbert Mead*, Chicago and London: University of Chicago Press.

Mead, G. H. (1927/1982), 1927 class lectures in social psychology, in Miller, D. L. (1982) (ed.), *The Individual and the Social Self*, Chicago and London: The University of Chicago Press.

Mead, G. H. (1934), *Mind, Self and Society*, Chicago: University of Chicago Press.

Mead, G. H. (1936), *Movements of Thought in the Nineteenth Century*, Chicago and London: University of Chicago Press.

Moscovici, S. (1976), *Social Influence and Social Change*, London and New York: Academic Press.

Moscovici, S. (1984), "The phenomenon of social representations," in Farr, F. M. and Moscovici, S., *Social Representations*, Cambridge: Cambridge University Press.

Nicolson, M. H. (1950), *The Breaking of the Circle*, Evanston: Northwestern University Press.

Piaget, J. (1953), *The Origin of Intelligence in the Child*, trs. Cook, M., London: Routledge and Kegan Paul.

Pylyshyn, Z. W. (1980), "Computation and cognition: issues in the foundations of cognitive science ," *Behavioral and Brain Sciences*, 3, 111–69.

Russell, J. (1978), *The Acquisiton of Knowledge*, New York: St Martin's Press.

Simon, H. (1962), "The architecture of complexity," *Proceedings of the American Philosophical Society*, 106, 467–82.

Sinclair, J. McH., and Coulhart, M. (1975), *Towards an Analysis of Discourse*, London: Oxford University Press.

von Cranach, M. (1982), "The psychological study of goal-directed action," in von Cranach, M. and Harre, R. (eds), *The Analysis of Action*, Cambridge and London: Cambridge University Press, and Paris: Editions de la Maison des Sciences de l'Homme.

Part V Culture and Causes of Development

10 The child's theory of mind and its cultural context

Paul Harris

The child's theory of mind and its cultural context

For the past few years, I have been studying children's ideas about emotion. I have not been looking directly at children's emotional development but rather at the way that they make sense of their own emotions or those of other people. This research is part of a much larger enterprise. Various investigators are currently examining several key features of the child's psychological understanding (Astington, Harris and Olson, 1988). For example, investigators are looking at children's understanding of beliefs (Perner, Leekam and Wimmer, 1987), of pretence and imagination (Leslie, 1988; Estes, Wellman and Woolley, in press), of visual experience (Flavell, 1986; Taylor, 1988) and of intention (Poulin-Dubois and Schultz, 1988).

This line of research sounds like what used to be called metacognition. There are, however, several important differences. First, it is being carried out with much younger children, typically 3- to 6-year-olds. Second, instead of being asked about school-like tasks such as learning a list of words or understanding a prose passage, children are questioned about events that might crop up in their daily life. For example, they are questioned about someone trying to see or find an object, about someone who does something by accident, or about someone who wants to hide their feelings.

These two changes reflect a fundamental reorientation. The earlier research on metacognition depicted the child as occasionally becoming self-conscious in the course of some other task, such as comprehension or retrieval. The child was depicted as an unselfconscious creature whose cognitive engine was usually left to run unattached. Only when there was some obstacle to comprehension or retrieval would the child reluctantly come to a halt, open the bonnet, and peer at the mysterious and complicated machinery inside.

At the risk of simplification, it is convenient to distinguish between two aspects of most mental events. On the one hand, there are the processes by which those mental events are generated, processes which do indeed remain hidden from conscious awareness. There are also the conscious products of those processes. Visual perception offers a fairly clear example of this distinction. Under normal circumstances, we can report on the conscious products of our visual experience. We can describe what we see, and how clearly we see it. Most of us, however, know very little about the process by which light falling on the retina activates signals in the brain that eventually result in our conscious visual experience. We do, admittedly, come to appreciate the conditions under which our visual experience of an object will be fragmentary, misleading or even prevented. Most of our knowledge, however, pertains to the disposition of the objects that we want to look at, or to our position relative to those objects. We know, in short, how to get a good view of something, but not how that view is mediated by the retina and the brain.

A similar story can be told with respect to other mental domains. We can report on our emotions, our intentions, and on the products of our imagination. We may also know the conditions under which we feel certain emotions, stick to our intentions, or let loose our imagination. Yet the neuropsychological processes that underlie our consciously experienced emotions, intentions and fantasies remain mysterious.

On the basis of the new research that I describe below, it seems likely that young children have quite a sophisticated theory of mind in so far as they possess considerable awareness of these conscious mental products. They can report some of the phenomenal regularities that adults know about and they appear to do so with little explicit adult tuition (Johnson, 1988). They are less accurate in their understanding of the conditions under which these conscious experiences will or will not arise, but their own experience is probably sufficient to allow them to discover those conditions.

I shall extrapolate well beyond these data, and argue for a position that was first sketched by Margaret Mead (1932) after her fieldwork in Manus. Her work was especially concerned with children's ideas about causality. She found that Manus children initially adopt a straightforward notion of cause and effect similar to that of the Western child or adult. Only later do they adopt the belief in ghosts, magical charms and animistic forces that is so prevalent among the adults of Manus. For example, if the Manus children were asked why a canoe went adrift, they emphasized the fact that it had not been properly moored. Their elders, on the other hand, would attribute evil intentions to the drifting canoe, even if they explained the immediate cause of its going adrift in the same way as the children. My guess is that this developmental pattern is equally applicable to children's ideas about the mind. Irrespective of culture, children will at first adopt similar ideas about the mind. Later they will begin to encounter the practices and theories prevalent in their culture. In the first place, they will have experiences that depend critically on the values, assumptions and institutions of their culture. As they make sense of such experiences, they will come to adopt a working model of the mind which necessarily reflects some of the cultural specificity of those experiences. Second, the mind, being an essentially metaphysical and mysterious entity, is conceived in a variety of ways. Different cultures have explored different options especially when they seek to explain behavior and experience that departs from the normal range (Heelas and Lock, 1981). The child who absorbs these ideas will eventually adopt a meta-theory: a set of culturally specific beliefs about the origins, dangerousness and controllability, of particular types of behavior and experience. I shall spell out my position by making a series of claims. I shall then attempt to substantiate each claim with empirical evidence where available.

A theory of mind

Children rapidly adopt a conception of mind that in certain respects resembles a theory. First, they invoke hidden causes to predict and explain the overt behavior of other people. Second, they interconnect those hidden causes, one with another; thus, the child can appreciate that a given action or emotion may need to be predicted or explained by taking two key mental constituents into account: beliefs and desires. Third, they impose the theory with minimal perceptual input. Children impose it on dolls and toy soldiers in their symbolic play; they impose it on animals and occasionally on inanimate entities.

A universal conception

The child's conception of mind is probably universal in the early years. Although cultures differ quite sharply in the ideas or collective representations that they have about the mind, children everywhere will have certain common experiences and arrive at a core set of conclusions.

The role of conscious experience

The starting point for the child's conception of mind is the existence of certain regularities and distinctions in conscious experience. Thus, perceptual experiences arise under specifiable conditions; the same is true for beliefs and emotions. Armed with immediate access to those experiences, the child does not observe other people and deduce or infer that they must be sad or acting on the basis of a given belief. Rather, the child projects onto other people the kinds of experiences that he or she would have given the surrounding circumstances. The prediction of other people's behavior is, in this respect, similar to the child's planning of his or her own behavior. Each depends on the imaginative projection of the self into a hypothetical but realizable set of circumstances.

Participation in social life

An understanding of other people is crucial for the child's participation in social life. Cultures often mark the emergence of that understanding by drawing the child into its activities; the child is increasingly offered new roles and more responsibilities. In the absence of that understanding, participation in social life is jeopardized. The autistic child offers an illustrative case. Recent evidence strongly suggests that autistic children are deficient in the ability to engage in the imaginative projection that is necessary for understanding other people and for planning their own lives. This quite specific deficit may explain the autistic child's tendency to withdraw from social life rather to participate.

Participation in the culture

Once children begin to participate in the social life of their culture, they will begin to have experiences whose meaning is constituted by certain assumptions and values that are special to that culture. These experiences will gradually enable the child to gain an understanding of the mind that is culture specific. The phenomenal regularities that the child begins to experience at this point will reflect the values and practices of the surrounding community.

Indigenous psychological theories

All cultures are confronted by certain problematic experiences and behaviors. Thus, each culture must deal with sick, mad, criminal or antisocial individuals whose behavior does not fit with normal expectations and does not yield to normal prescriptions. Similarly, there are certain experiences – dreams, hallucinations, intense feelings of depression or anger – whose quality deviates from our normal mental life. It is particularly with respect to these deviations from the normal that cultures will begin to elaborate meta-theories in an effort to explain, predict and control them.

In summary, the child quickly elaborates a theory-like conception of mind that reflects certain regularities of experience, and will probably differ very little from one culture to another. This theory is a precondition for normal participation in the culture. As that participation becomes more extensive and elaborate the child will begin to draw conclusions – again based on personal experience – which increasingly reflect the institutions and practices of the surrounding community. Mostly as an observer, but sometimes as a patient, the child will also learn how the culture explains and handles abnormal, psychological experiences. To the initial, core theory, a culturally specific meta-theory will be added.

Evidence and elaboration

A theory of mind

I gave three indices of the theory-like nature of the child's theory of mind. The postulation of hidden entities, of interconnectedness, and the need for minimal perceptual input to drive the theory. One illustration of the child's postulation of hidden entities comes from my own work on children's understanding of the distinction between real and apparent emotion. By the age of six years, children systematically appreciate that the emotion that someone really feels need not correspond to the emotion that the person displays. Thus, told about a story character who falls over, but tries to hide their real feelings for fear of being teased by the other children, 6-year-olds systematically judge that the story character will try to look positive or happy while really feeling sad (Harris and Gross, 1988; Harris, Donnelly, Guz and Pitt-Watson, 1986). Moreover, children know that such misleading displays can hide the protagonist's true feelings from onlookers. Asked to say what emotion other story characters would attribute to the child who fell over, 6-year-olds judge that they would mistakenly attribute positive rather than negative feelings (Gross and Harris, 1988).

The understanding of interconnectedness can be illustrated from the same area of research. Consider a child who has been tricked. She is given a box of Smarties, but her favorite sweets are Polos not Smarties; unbeknown to her, however, the normal contents – Smarties – have been replaced by her favorite – Polos. How will she feel when first given the container before opening it? And how will she feel after opening it? The majority of 6-year-olds correctly answer *sad* to the first question and *happy* to the second (Harris, Johnson, Hutton and Andrews, 1989). Notice that the answers to these questions depend on the joint appreciation of the girl's beliefs and desires: to answer the first question correctly, the child must take into account her false belief about the contents and her lack of desire for the supposed contents; to answer the second question, the child must take into account her revised belief about the contents and her desire for those contents. Thus, these results illustrate how the child can coordinate two crucial components of the standard folk theory of the mind – beliefs and desires – in order to predict a third component, namely emotion.

The need for minimal perceptual input is best illustrated by children's symbolic play. As Leslie (1987) has pointed out the emergence of symbolic play is early and rapid – at around eighteen months in normal children. During such play, children impose on inanimate objects such as dolls and toy soldiers, the full panoply of theoretical entities that they deploy in their interaction with human beings. They attribute intentions, desires, perceptual experiences and emotions to their play-things (Bretherton, 1984). Moreover, children make such attributions safe in the knowledge that their dolls and toys do not really experience these mental states (Gelman, Spelke and Meck, 1984). It is a question of deliberate make-believe not epistemological confusion.

A universal conception

To what extent are the attainments that I have described a universal acquisition? There is limited data to answer this question at present, but all of it is encouraging. A considerable body of evidence indicates that even 4-year-olds can make allowances for the fact that someone may not have seen an object re-positioned so that they will falsely believe it to still be in its old position and search for it there. This is a major achievement because it shows that the child can distinguish between his or her own (true) beliefs about the object's location and the temporarily mistaken beliefs that someone else might entertain. The false-belief task has been carried out in Austria, the United Kingdom and the United States. In each culture, the data reveal a marked shift between three and five years of age. Children of 3–4 years typically make the wrong prediction. They predict that the person will look in the position where they themselves know the object to be really hidden. Children of 4–5 years, on the other hand, correctly anticipate that the story character will act on his or her false belief (Perner *et al.*, 1987).

Flavell (1986) has examined a conceptually related task: the case of misleading appearances. The child is confronted with a fake object (such as a sponge that has been painted to look like a rock) and allowed to explore it to discover its true identity. The child is then asked to say both what it looks like and what it really is. Again a marked shift is observed between three and five years of age. The older child can systematically distinguish between the object's real identity and its apparent identity, whereas the younger children tend to focus on one to the exclusion of the other. This task is conceptually related to the false-belief task, since it requires the child to bear in mind the distinction between reality on the one hand, and the beliefs that someone might entertain about that reality given limited or abnormal perceptual access to it. Performance on the appearance–reality task also shows cross-cultural stability. Flavell, Zhang, Zou, Dong, and Qi (1983) tested 3- to 5-year-olds in mainland China and found the same age change that they had observed in the United States.

The distinction between real and apparent emotion also emerges at approximately the same age in different cultures. Gardner, Harris, Ohmoto and Hamazaki, (1988) tested 4- and 6-

year-olds in Japan using stories similar to those already employed with children from the United States and the United Kingdom. Results across the three cultures were quite comparable, with 4-year-olds having trouble realizing that the emotion displayed might not correspond to the real emotion and 6-year-olds systematically making the distinction.

All of these results support the claim that the same theory of mind emerges universally in the young child with approximately the same timetable. The data are encouraging but scarcely overwhelming at this stage. First, although the cultures that have been compared are quite disparate in being drawn from both East and West, the children may have been exposed to certain similar experiences. All had been to school or preschool; all lived in industrialized, or semi-industrialized communities with a fairly high literacy rate. We need to know how children in pre-literate, pre-industrial cultures perform, particularly in cultures whose adult members have explicitly different concepts of the mind from those normally held in the West.

The role of conscious experience
I have argued that the child starts off building a theory of mind on the basis of immediate conscious experience. Evidence for this claim is necessarily indirect and tentative, but it is beginning to accumulate. Consider the time-course of emotion. As adults, we know that emotion varies markedly in intensity and that typically we feel most intensely about an emotionally-charged event immediately after it has happened. Thereafter, we experience a gradual waning in the intensity of the emotion. Do young children know this? A set of experiments now indicates that young children have a good grasp of this relationship between time and intensity. Thus, they judge that immediately after an emotionally-charged event such as the death of a pet, or the gift of a bicycle, someone will feel intense emotion. Asked to say how that person will feel later on that day or the following day, they appropriately judge that feelings will wane in intensity from one period to the next. These judgments are made by 10-, 6-, and even 4-year-olds and they are made by children in the West and by children in China (Harris, 1983; Harris, Guz, Lipian and Man-Shu, 1985; Taylor and Harris, 1983).

How do young children work out this relationship between time and intensity? I would argue that they experience it for themselves, and they assume that other people will experience the same relationship when they feel an emotion. Other interpretations are possible, however. For example, children might observe other people and notice that signs of affect gradually wane in intensity as time passes by. Yet, tracing back the signs of emotion in such circumstances would be difficult. How is the child to connect up the intermittent signs of emotion in other people and to see them as part of a coherent, causal chain?

The strongest evidence for the role of phenomenal experience comes from some recent experiments by Estes, Wellman and Woolley (in press). They have examined a crucial aspect of the children's theory of mind: their appreciation that one can conjure up mental images of an object. Preschoolers understand that such mental entities are quite different from the real object that they represent. Real objects can be touched and can be seen by other people whereas mental images cannot be touched or seen by others. On the other hand, mental images can be transformed at will: just by thinking about it one can transform an image of a cup so that the cup appears upside down; mere thought will have no effect on the real cup, however. Five-year-olds appropriately distinguished the mental entity and the real entity with little prompting; 3- and 4-year-olds were also quite systematic, but they were less optimistic than the older children about the possibility of transforming their mental image (e.g. imagining an inverted cup) just by thinking about it. Wellman and his colleagues went on to ask these young skeptics to try, nevertheless, to transform their image of the cup. Afterwards, the majority agreed that it was possible. This was not simply a case of acquiescence because another request by the experimenter (to transform a cup hidden in a box just by thinking) did not lead children to change their initial claim that this was an impossible task. This pattern provides strong evidence, not just that young children have mobile visual imagery, but that they can introspectively attend to it, note its operation, and reach a new conclusion about their mental capacities.

I have discussed two examples where it seems plausible to suppose that children learn from immediate introspective awareness with little explicit instruction from adults. Other results do not fit this interpretation so readily. Take for example, the way in which young children predict other people's actions or emotions by taking into account their false beliefs. There is no particular introspective quality associated with false beliefs. Moreover, if children were to project their own beliefs onto the characters, they would incorrectly attribute true beliefs to them. How then do children solve the problem? One possible answer is that children come to use a combination of imagination and introspection (Johnson, 1988). Consider the false-belief task in more detail for a moment. The child observes one story character deposit some chocolate in a box and leave. A second character arrives, moves the chocolate to a new location and leaves. The first character returns and is about to look for the chocolate. The child is then posed the critical question: where will the character search? Older children of 4–5 years correctly predict that the character will look where the chocolate was originally deposited whereas younger children of 3–4 years predict that the story character will go to the new location (Perner *et al.*, 1987). To make correct predictions, children must clearly take into account the fact that the character holds a false belief about the object's location. How do children work this out? One possibility is that they simply imagine what they would do were they in the shoes of the first character. Specifically, they imagine themselves outside of the room when the transfer of the chocolate takes place, and therefore unable to see it. They imagine themselves coming back to the room having originally put the chocolate in a particular box. Imagining themselves in this role, they can mentally simulate the visual experience, the consequent beliefs, and the eventual actions of the character. Thus, the false belief problem can be solved by introspection provided the child is capable of engaging in the kind of imaginative simulation that I have described (Harris, 1989).

When spelled out so explicitly, it seems unlikely that children would be capable of such role-playing. On the other hand, as I noted earlier, children's symbolic play exhibits an enormous amount of imaginative simulation. The child takes a toy character, places that character in some hypothetical albeit familiar situation, such as having a bath or going shopping, and proceeds to generate various potential experiences that could follow from being in those situations. The bath water feels too hot, and the character is made to jump out of the bath. The character wants to buy something that the shopkeeper does not have and leaves the shop. Note that even these two prosaic examples require that the child conjure up counterfactual states embedded within the play-setting and of course the play-setting itself depends on various counterfactual assumptions. Thus, each play character is invested with a desire for some counterfactual state in the course of their activity, and more generally, the play characters are invested with thoughts, desires, and experiences that inanimate toys cannot really have.

Participation in social life

To begin to appreciate the importance of the child's theory of mind for everyday social life, we may briefly examine the disorder of childhood autism. Some of the key symptoms of autism are an inability to form satisfactory relationships with other people, an obsessive preoccupation with certain routines and activities, and a lack of spontaneous play, especially pretend play (Kanner, 1943).

Leslie (1987) has recently offered a new interpretation of these symptoms. He argues that unlike normal children, autistic children have considerable difficulty in imagining some counterfactual state of affairs. An initial piece of evidence for this claim came from a replication of the false-belief task with autistic children. The customary test situation involving two story characters and a piece of chocolate which is deposited by one and moved by the second was enacted for a group of autistic children. The majority failed to take the first story character's false belief into account and predicted that on her return she would search in the new location where the chocolate was currently hidden. Comparison groups of normal and Down's syndrome children whose average IQ was actually lower than that of the autistic children performed much better (Baron-Cohen, Leslie and Frith, 1985).

Two further experiments have elaborated this basic finding. Autistic children do not have a general difficulty with causal reasoning. Given a jumbled set of pictures about a purely mechanical sequence (e.g. a man kicks a boulder; it rolls downhill; and lands in the water), they readily re-ordered the pictures. On the other hand, given a jumbled set of pictures about a psychological sequence (e.g. a girl puts down her teddy, and turns to pick a flower; a boy surreptitiously removes the teddy; she turns around to find it missing), the autistic children did very poorly. Moreover, when asked to describe the psychological sequence, they tended to stick to a surface description of behavior, with little reference to the beliefs, motives or emotional reactions of the story characters (Baron-Cohen, Leslie and Frith, 1986).

In a final study, Baron-Cohen (1987) has established that autistic children's play is especially deficient in pretend or make-believe play. This is exactly what would be expected if children cannot imagine counterfactual states of affairs: they should not be able to engage in pretend, and they should not be able to endow story characters with beliefs about counterfactual states of affairs which the characters mistakenly take to be true.

Taken together, these results suggest that the autistic child has a very limited understanding of other people's mental states. That limited understanding may well explain the autistic child's difficulties in participating in everyday social life. We can also plausibly trace that limited understanding back to one of the major symptoms: the absence of play, especially pretend play. Such play appears to reflect the kind of imaginative simulation which is vital for understanding other people. Without such a capacity for simulation, other people remain unpredictable, and indeed inalienably different from oneself.

Finally, we may have a tentative explanation for the autistic child's rigid preoccupation with routine and sameness. Suppose that autistic children have difficulty in envisaging a state of affairs that does not obtain. In particular, suppose that they have difficulty in imaginatively simulating various courses of action that they might take and in rehearsing their emotional reactions to those imaginery possibilities. If that were true, and it is a highly plausible consequence of the kind of deficits that have been described above then change would not fit into a set of expectations about how things might proceed differently from the way they normally do. Instead, change would be a total, unexpected, and unrehearsed departure from what is normally the case. Accordingly, such departures might well be much more threatening to the autistic child than to the normal child.

Participation in the culture

As children begin to participate in their culture, and to take on roles and responsibilities, they will inevitably begin to have experiences which are not universals of childhood, but which are imbued with the values and assumptions of their community. These experiences will in turn shape the way in which the child conceives of mental processes. Two recent investigations illustrate this theme. Harris, Olthof, Meerum Terwogt and Hardman (1987) presented children with a set of emotion terms – *happy, sad, disappointed, relieved, proud,* and so forth. The children were asked to supply for each term a situation that would be likely to provoke that emotion. The study was carried out in Amsterdam, Oxford, and Pangma. Pangma is a tiny village in the Eastern Himalayas of Nepal. The children in this village belong to the Rai tribe and were interviewed in their native language Lohorung, a language of Tibeto-Burmese descent. The children of Pangma grow up in a radically different culture from their peers in Oxford or Amsterdam. They have never seen a car, a bicycle or a road. There is no television in the village and radios are scarce. The children have some schooling, but they also have to help their parents by looking after cattle or working in the fields or taking care of their younger siblings. Like children in many Third World countries, they participate responsibly in the adult culture more rapidly and more fully than their Western counterparts. (For more ethnographic information see Hardman, 1981.)

If one looks at the replies of the Lohorung children, it is clear that their participation has had an impact. In describing the situations that would make them *angry* or *sad*, they frequently pick on situations that are specific to the culture, or which are invested with a particular significance by the culture. For example, children who looked after the cattle (almost invariably boys) often mentioned situations that would arise during their vigil. It turns out that watching cattle is a variegated, and emotionally-charged experience for them. When we asked them: "What would make you *angry?*", a typical reply from a 14-year-old was as follows: "If the cattle go into the maize or other crops and eat what you've grown, then you have to run after them and shout at them and then you're angry." Replies to other questions elaborated on this theme: "When would you feel *pity?*" – "When the cattle go into the fields and eat the crops, then you feel pity for the crop-owners" explained an 11-year-old boy. "When would you feel *sad?*" – "When I see other cattle eating my father's crops," said a 12-year-old boy. "When would you feel *relieved?*" – "If somebody's cattle ate all the crops but then the owner of the cattle pays it back, then you feel relief" said an 8-year-old boy.

Going to school was another theme in which the children's

definition of the situation was obviously culture-specific. Since they are frequently expected to help their parents, the opportunity to go to school is seen as a privilege rather than as an irksome obligation. Accordingly, not being sent to school was cited in connection with sadness not happiness. Conversely, an older brother who was studying when you were not was cited in connection with jealousy rather than relief.

Despite their positive evaluation of school, the replies of the Lohorung children were more obviously concerned with the misfortunes and tragedies of the adult world. Not infrequently, they spoke of illness and death, of work and its mishaps, of threatened livelihoods and poverty. By contrast, the replies of the Dutch and English children that we interviewed suggested a more safe and sanitized world of school, pets and toys.

How should we interpret these results? The most conservative interpretation is that the child's culture has an impact only on the conditions in which particular emotions are experienced not on the intrinsic nature of those emotions. Thus, a Lohorung boy who feels *sad* when he sees cattle eating his father's crops might feel the same emotion as his Western counterpart who loses all his marbles. According to this line of argument culture would simply take a fixed set of emotions and attach them to culturally specific eliciting conditions. Cultures would differ only in the way that particular situations are appraised but not in the emotions that ensue following that appraisal. Theorists who have argued for a set of discrete, universal emotions usually make this minimal concession to the impact of culture (Ekman, 1973).

A second and less conservative interpretation is that as a result of becoming linked to a particular set of eliciting conditions, the emotions themselves become differentially appraised by the culture; consequently, the emotion produces a qualitatively different experience, since its causal impact on other emotions becomes orchestrated by the culture. For example, among the Lohorung it is expected that a child should be shy among older male relatives such as an older brother or uncle. Shyness is regarded, to some extent, as a mark of respect. Geertz (1959) reports a similar pattern among Javanese children. Since the child is expected to feel shy, and since that shyness is approved, it is likely that the Lohorung child experiences shyness quite differently from his or her Western peer. Shyness is not something that invites further feelings of shame or anxiety; it is something that the child might even feel proud to show since it will elicit approval from adults. This brief example illustrates the way in which culture can go beyond establishing the conditions for the elicitation of an emotion. Assuming that there are initially discrete emotions, it can, in principle, shift the relations among those emotions, placing some in close causal conjunction and dissociating others.

Some anthropologists, however, have argued for a third, more radical possibility. They claim that culture does not simply alter the causal conditions for a set of preexisting emotions, it manufactures new culturally-specific emotions. This is essentially the position taken by ethnographers such as Lutz (1987) and Rosaldo (1980). Briefly, they assert that certain emotions are not simply familiar emotions being aroused in culturally-specific circumstances. Rather the emotion itself is defined and constituted by participation in particular cultural frames. Like "gold-fever" or "machismo" or "*gemütlichkeit*", the emotion presupposes a set of cultural assumptions before it can ever be experienced. Our data do not allow us to rule out this third possibility. If correct, it would suggest that at certain points, the child's understanding of emotion will be almost indissolubly connected with an understanding of the culture as a whole.

The second study was carried out in an altogether different culture. Here, young males do not help with adult tasks in any way. Instead, they are separated from their parents, sometimes from eight years of age upward and sent off to be trained in large groups sub-divided into cohorts of the same chronological age. The boys eat, sleep, study and worship in isolation from the rest of the community. The culture that I have in mind is that of Britain where children, particularly boys, are often sent to boarding school.

We interviewed 8- and 13-year-old boys who had recently arrived at a new boarding schoool about their emotional reactions to the new school and to being separated from their parents (Harris, 1989). One of the most intriguing findings emerged when we asked the boys the following question:

"Say you were missing your friends and family back home and wanted to cheer yourself up. Is there anything you could do to cheer yourself up?"

The bulk of the coping strategies proposed by the boys fell into two quite distinct categories. One category we called *Contact with Home*. The boys mentioned various ways in which they could reduce the psychological distance between themselves and their home. For example, they might say: "I look at things I've brought from home", or "Draw a picture and send to to your Mummy and Daddy", or "If you brought your teddy from home, it helps" or "Look at photographs of your parents".

The second category we called *Distraction*. The boys described ways in which they could distract themselves from the distress of separation. For example, they might say: "If there was a film on that night, you could go and watch it and distract yourself" or "I could try and think about something else ... the time felt so long, so what I did was to try and fit everything in, try to understand everything – like maybe pass all of my tests and stuff like that" or "When I first came here, I used to think out problems – so that your whole mind is centered on those problems ... Sometimes, when I felt a bit homesick, I started doing mathematical problems."

Our results show that the *Distraction* strategy is the one that most of the boys favor, particularly among the 8-year-olds who were experiencing boarding for the first time. Since the boys cannot actually go home – they can only write letters or look at photographs – the anesthetic of *Contact with Home* is necessarily short-lived. One 13-year-old explained the advantages and disadvantages of the two strategies as follows: "The worst thing to do is ring home because the time comes when you have to say goodbye. I'd get into things ... Hopefully, I could make myself forget, keep myself occupied the whole time." Thus, *Distraction* provides a more permanent solution, although as some boys pointed out, that strategy cannot always be put into effect. You can keep your mind occupied during the day but, at night before you go to sleep, you often start thinking about your home, and may become tearful as a result.

We cannot prove it from our interview data, but it is reasonable to suppose that the institutional framework within which separation is experienced helps the boys to learn about the effectiveness of the *Distraction* strategy. The boarding school offers its pupils a highly organized schedule of collective activities: lessons, games, religious worship, sport, meetings and meals.

These activities provide opportunities for distraction throughout most of the child's waking hours. As a result, boys in boarding school have learned from personal experience about its benefits.

Indirect evidence for this interpretation comes from a study of children in hospital (Harris and Lipian, 1989; Lipian, 1985). When these children were asked about ways in which they could cheer themselves up, they rarely mentioned the use of the *Distraction* strategy. Like the boarding school, the hospital has a routine, and it is sometimes a fairly rigid one, but it is an intermittent and less distracting schedule than that of the boarding school. Thus, children in hospital sometimes experience long periods of boredom or inactivity and will have fewer opportunities to discover the benefits of *Distraction*.

Indigenous psychological theories

In the previous section, I discussed ways in which the practices of the culture would lead children to adopt increasingly divergent theories of the mind, consonant with the personal experiences that they had had in that culture. Thus, the boarding schoolboy learns about the way in which loss and homesickness can be anesthetized by distraction. However, cultures differ not only in their practices but also in their theories about the way that the mind works. Such theories may be more or less explicitly conveyed to the child at various points in development and they may even contradict conclusions that the child has arrived at on independent grounds. By way of illustration, I will consider the example of dreams.

Piaget (1929) and more recently Kohlberg (1966) investigated children's ideas about dreams. They both claimed that children start off by confusing dreams with reality but gradually acknowledge at about eight or nine years that dreams are unreal, internal, mental processes. This timetable is obviously quite slow compared to the one proposed by Estes *et al.* (in press) for children's understanding of mental images. Part of the explanation for this difference may be that both Piaget and Kohlberg posed various misleading questions (e.g. "Where do dreams come from?" and "What are they made of?") which imply that dreams, like physical objects, are made of a particular substance and come from or can occupy a particular place. However, dreams take place during sleep; they appear real during sleep; they quite often recur quite involuntarily; and they sometimes provoke strong emotion.

Accordingly, they may be much more easily confused with reality than the more fleeting fantasies or images that can be deliberately conjured up during everyday waking activities. Thus, although further research may show that children identify dreams as purely subjective, mental phenomena at an earlier age than Piaget and Kohlberg suggested, it seems unlikely that the child's understanding of dreams develops in tandem with his or her understanding of mental images.[1] In any case, even if we adopt the conservative timetable proposed by Piaget and Kohlberg, we still arrive at the conclusion that children appreciate that dreams are purely mental phenomena long before they reach adulthood.

This conclusion is quite striking when we examine the conception of dreaming that is prevalent in certain preliterate cultures. Consider, for example, the model of dreams espoused by the Lohorung, introduced in an earlier section (Hardman, 1981). They postulate the existence of "lawa" in each person, which we may roughly translate as spirit or awareness. The child's is held to be quite timorous, and especially when the child is frightened, may leave the child's body altogether. A person's "lawa" also leaves his or her body during sleep; untrammelled by the constraints of time and space, it is free to wander. It may go to a far off country, to the spirit world or even to another time located in the past or the future. Thus, for the Lohorung, dreams are readily explained: they reflect the wandering of one's "lawa" as one sleeps. Moreover, since "lawa" sometimes visits the future, dreams may indicate what the future holds.

In some respects, the Lohorung regard dream events as real events, although these events take place in another metaphysical realm. Given that status, they can serve as clues to the future, particularly if they are interpreted by someone skilled at dream interpretation. Admittedly, some adults in the West also take dreams to be a prognostication or warning about the future, just as they did in seventeenth-century England (Thomas, 1971) but such claims deviate from the standard contemporary model and are usually met with varying degrees of skepticism.

The notion of a wandering spirit is found in other preliterate cultures. For example, Tonkinson (1970) reports a similar conception of dreams in his ethnography of an Australian aboriginal group. The Atayal, a Malaysian aboriginal group on Taiwan, also believe that the soul leaves the body during dreams and experiences things in far away places (Kohlberg, 1966).

Given this disagreement between the Western child and the adult Lohorung, or Atayal, it is tempting to conclude that the Western child arrives at the conclusion that dreams are purely subjective, mental experiences because he or she is told exactly that by reassuring adults, particularly after a bad dream or nightmare. One way to assess this interpretation is to examine the concept of dreams adopted by children in a preliterate society. If children adopt the conception that is offered to them by adults, we should find that among the Lohorung or the Atayal, children never believe that dreams are purely unreal, mental phenomena. Instead, they should gradually come to assert that the scenes that they observe during their dreams are scenes in a different time or place being witnessed by their soul or "lawa"

Kohlberg (1966) carried out such an investigation among the Atayal. Strikingly, he found that until early adolescence, children moved through the same conceptual stages as their Western peers, albeit somewhat more slowly. thus, up to the age of 11 years, Atayal children increasingly acknowledged that dreams are unreal, internal, mental phenomena. Older children, by contrast began to deny that dreams were mental and internal. This is, of course, exactly the pattern that we would expect if the core theory of mental phenomena is universal, but is gradually decked out with various, culture-specific, meta-theories.

Conclusions

Our current view of cognitive development remains heavily influenced by Piaget. He argues that the young child is misled by appearances and only gradually constructs, regardless of culture, a theory that acknowledges the underlying stability and predictability of the world. This is a profound and also a seductive vision of the nature of cognitive development.

The study of the child's psychological knowledge shows that it is also deeply misleading. It is misleading in three important respects. First, in explaining and predicting other people's behavior the child imposes a theory on the observable data and does so quite rapidly. That theory is not constructed by deduction or disequilibration. It appears to be the joint product of conscious awareness and imaginative projection. Second, although the young child is, no doubt, misled by surface appearances, he or she can penetrate or override perceptual appearances with an act of the imagination. Piaget was inclined to see such imaginative play as somewhat disreputable from a strictly cognitive point of view. He assumed that during play reality was subordinated to and even distorted by the child's own schemata. Yet the autistic child stays close to surface reality, and has a good grasp of the mechanical universe; the fact that the autistic child often fails to grasp the psychological meaning that underlies the social universe, illustrates how cognitive development is impoverished not accelerated by the suppression of the imagination. Third, the later refinement of the child's theory of mind does not depend on the construction of a set of pan-cultural principles. Human behavior is too subject to the vagaries of cultural practice to be approached in such a spirit. Such practices are best understood by increased participation. It is experience that the child lacks, not intellectual penetration.

Acknowledgments

This chapter has benefitted from discussion with various friends and colleagues, discussion that in some cases has extended over several years. I would particularly like to thank Paul Heelas and Charlotte Hardman who helped me to find my way on various anthropological expeditions; Carl Johnson who convinced me that I was taking some ideas too seriously and others not seriously enough; and Henry Wellman who repeatedly demonstrated the capacities of preschool children to me.

Note

1. Henry Wellman (personal communication) has gathered data on the understanding of dreams by 3- to 5 year-old children. Preliminary analysis indicates that 5-year-olds systematically appreciate the mental origin of dreams, whereas younger children often assume an external source.

References

Astington, J. W., Harris, P. L. and Olson, D. R. (1988), *Developing Theories of Mind*, Cambridge: Cambridge University Press.

Baron-Cohen, S., Leslie, A. M. and Frith, U. (1985), "Does the autistic child have a theory of mind?" *Cognition*, 21, 37–46.

Baron-Cohen, S., Leslie, A. M. and Frith, U. (1986), "Mechanical behavioural and intentional understanding of picture stories in autistic children," *British Journal of Developmental Psychology*, 4, 113–25.

Baron-Cohen, S. (1987), "Autism and symbolic play," *British Journal of Developmental Psychology*, 5, 139–48.

Bretherton, I. (1984), *Symbolic Play*, Orlando, Fla: Academic Press.

Ekman, P. (1973), *Darwin and Facial Expression: A Century of Research in Review*, New York: Academic Press.

Estes, D., Wellman, H. M. and Woolley, J. D. (in press), "Children's understanding of mental phenomena," *Advances in Child Development*, Orlando, Fla: Academic Press.

Flavell, J. (1986), "The development of children's knowledge about the appearance-reality distinction," *American Psychologist*, 41, 418–25.

Flavell, J. H., Zhang, X-D., Zou, H., Dong, Q. and Qi, S. (1983), "A comparison of the appearance-reality distinction in the People's Republic of China and the United States," *Cognitive Psychology*, 15, 459–66.

Gardner, D., Harris, P. L., Ohmoto, M. and Hamazaki, T. (1988), "Understanding of the distinction between real and apparent emotion by Japanese children," *International Journal of Behavioral Development*, 11, 203–18.

Geertz, H. (1959), "The vocabulary of emotion: a study in Javanese socialization processes," *Psychiatry*, 22, 225–37.

Gelman, R., Spelke, E. S. and Meck, E. (1989), "What preschoolers know about animate and inanimate objects," in Sloboda, J., Rogers, D., Bryant, P. E. and Cramer, R. (eds), *The Acquisition of Symbolic Skills*, pp. 297–326, London: Plenum.

Gross, D. and Harris, P. L. (1988), "Understanding false beliefs about emotion," *International Journal of Behavioral Development*, 11, 475–88.

Hardman, C. E. (1981), "The psychology of conformity and self-expression among the Lohorung Rai of East Nepal," in Heelas, P. and Lock, A. (eds), *Indigenous Psychologies*, London: Academic Press.

Harris, P. L. (1983), "Children's understanding of the link between situation and emotion," *Journal of Experimental Child Psychology*, 36, 490–509.

Harris, P. L. (1989), *Children and Emotion: The Development of Psychological Understanding*, Oxford: Blackwell.

Harris, P. L., Johnson, C. N., Hutton, D. and Andrews, G. (1989), "Young children's theory of mind and emotion," *Cognition and Emotion*, 3.

Harris, P. L. and Gross, D. (1988), "Children's understanding of real and apparent emotion," in Astington, J. W., Harris, P. L. and Olson, D. R. (eds), *Developing Theories of Mind*, Cambridge: Cambridge University Press.

Harris, P. L., Donnelly, K., Guz, G. R. and Pitt-Watson, R. (1986), "Children's understanding of the distinction between real and apparent emotion," *Child Development*, 57, 895–909.

Harris, P. L., Guz, G. R., Lipian, M. S. and Man-Shu, Z. (1985), "Insight into the time-course of emotion among Western and Chinese Children," *Child Development*, 56, 972–88.

Harris, P. L., Olthof, T., Meerum Terwogt, M. and Hardman, C. E. (1987), "Children's knowledge of the situations that provoke emotion," *International Journal of Behavioral Development*, 10, 319–44.

Harris, P. L. and Lipian, M. S. (1989), "Understanding emotion and experiencing emotion," in Saarni, C. and Harris, P. L. (eds), *Children's Understanding of Emotion*, New York: Cambridge University Press.

Heelas, P. and Lock, A. (1981), *Indigenous Psychologies*, London: Academic Press.

Johnson, C. N. (1988), "Theory of mind and the structure of conscious experience," in Astington, J., Harris, P. L. and Olson, D. R. (eds), *Developing Theories of Mind*, Cambridge: Cambridge University Press.

Kanner, L. (1943), "Autistic disturbances of affective contact," *Nervous Child*, 2, 217–50.

Kohlberg, L. (1966), "Cognitive stages and preschool education," *Human Development*, 9, 5–17.

Leslie, A. M. (1987), "Pretence and representation: the origins of 'theory of mind'," *Psychological Review*, 94, 412–26.

Leslie, A. M. (1988), "Some implications of pretense for mechanisms underlying the child's theory of mind," in Astington, J. W., Harris, P. L. and Olson, D. R. (eds), *Developing Theories of Mind*, Cambridge: Cambridge University Press.

Lipian, M. S. (1985), "Ill-conceived feelings: Developing concepts of the emotions associated with illness in healthy and acutely ill children." Unpublished doctoral dissertation, Yale University.

Lutz, C. (1987), "Goals, events and understanding in Ifaluk emotion theory," in Holland, D. and Quinn, N. (eds), *Cultural Models in Language and Thought*, pp. 290–312, Cambridge: Cambridge University Press.

Perner, J., Leekam, S. and Wimmer, H. (1987), "Three-year-olds' difficulty in understanding false belief: cognitive limitation, lack of knowledge, or pragmatic misunderstanding?" *British Journal of Developmental Psychology*, 5, 125–37.

Piaget, J. (1929), *The Child's Conception of the World*, London: Routledge and Kegan Paul.

Mead, M. (1932), "An investigation of the thought of primitive children, with special reference to animism," *Journal of the Royal Anthropological Institute*, 62, 173–90.

Poulin-Dubois, D. and Schultz, T. R. (1988), "The development of the understanding of human behavior: From agency to intentionality," in Astington, J. W., Harris, P. L. and Olson, D. R. (eds), *Developing Theories of Mind*, Cambridge: Cambridge University Press.

Rosaldo, M. Z. (1980), *Knowledge and Passion*, Cambridge: Cambridge University Press.

Taylor, M. (1988), "The development of children's understanding of the seeing-knowing distinction," in Astington, J. W., Harris, P. L. and Olson, D. R. (eds), *Developing Theories of Mind*, Cambridge: Cambridge University Press.

Taylor, D. A., and Harris, P. L. (1983), "Knowledge of the link between emotion and memory among normal and maladjusted boys," *Developmental Psychology*, 19, 832–8.

Thomas, K. (1971), *Religion and the Decline of Magic*, London: Weidenfeld and Nicolson.

Tonkinson, R. (1970), "Aboriginal dream-spirit beliefs in a contact situation: Jigalong, Western Australia," in Berndt, R. M. (ed.), *Australian Aboriginal Anthropology*, Perth, Western Australia: Western Australia Press.

11 Life-history perspectives on human development

James S. Chisholm

Introduction

In spite of good intentions and enlightened protestations of interactionism, the control of ontogeny is still too often, at least implicitly, partitioned into genetic and environmental causes. As a result, understanding the organism–environment dialectics that underlie developmental processes remains elusive (e.g. Oyama, 1985). Based on Waddington's (1975) argument that the evolution of organisms must be understood as the evolution of developmental systems, the purpose of this chapter is to suggest that a fresh approach to understanding these dialectics may be found in the emerging field of life-history theory. Life-history theory is a combination of evolutionary ecology and demography that is especially concerned with the question of how organism–environment interactions may mediate the relationships between phylogeny and ontogeny. An overview of life-history theory and its central tenets will be followed by examples of how life-history theory is beginning to be used in anthropology.

The components and development of fitness

Life-history theory is emerging as an energetic area of study in large part because of a growing dissatisfaction with evolutionary biology's current emphasis on inclusive fitness theory. While it is generally recognized that the logic of inclusive fitness theory is sound, many are also recognizing that inclusive fitness theory is silent on the relationship of the genotype to the phenotype (e.g. Bateson, 1982). Since the relationship of the genotype to the phenotype can only be a developmental one, it follows that a natural science of behavior must be based on more than inclusive fitness theory.

The problem stems ultimately from the history of evolutionary theory. With the addition of genetics, population genetics, and demography to Darwin's theory of evolution by natural selection, the resulting New Synthesis was able to generate powerfully predictive and wonderfully elegant formal models of how gene pools change. As a consequence, the focus of evolutionary biology, and especially today's sociobiology, has been almost exclusively on the factors that change the frequency distributions in gene pools from one generation to the next. But because of their very success in this endeavor they have also neglected the fact that selection operates on the phenotype, not the genotype, and that it is the success of the phenotype that determines which genes are copied into the next generation. Perforce, they have also neglected the ecological interactions of the phenotype with the environment that influence fitness, and have neglected especially the organism-environment dialectics that *produce* the phenotype (e.g. Bateson, 1981, 1982; Bonner, 1982; Gould, 1977, 1982; Oyama, 1985; Stearns, 1982; Plotkin and Odling-Smee, 1979, 1981).

What we need now, as Stearns (1982, p. 254) puts it, is a "developmental evolutionary ecology" – a combination of evolutionary ecology, life-history theory, and developmental biology, all of which have a greater focus on the phenotype and its development than inclusive fitness theory. A major impetus for this sort of study has been the question: if selection has designed organisms to maximize fitness, and if fitness is defined as reproductive success (e.g. Daly and Wilson, 1983, p. 21), why don't all organisms reproduce as much as they can (as in r-selection)? The answer, of course, is that frequently total lifetime reproductive output can be maximized by adopting a slow and cautious reproductive strategy which includes high parental investment in a small number of high-quality offspring (as in K-selection). but it follows from this that fitness must be measured in terms of its *components*, for there is more to fitness than simple number of offspring, and natural selection cannot be expected to maximise reproductive output at the expense of adaptations for *survival* and the *optimal ontogenetic preparation for reproduction*. Thus, number of offspring, even surviving offspring, is not the only or necessarily the best measure of fitness. Instead, fitness may better be assessed by determining the optimal allocation of resources to the inherently conflicting demands of survival, optimal growth and development and other preparations for reproduction (and parenting), and reproductive output itself (e.g. Gadgil and Bossert, 1970; Horn and Rubenstein, 1984; Johnston, 1982; Krebs and Davies, 1984; Lack, 1954; MacArthur, 1962; MacArthur and Wilson, 1967; Pianka, 1970; Stearns, 1976, 1977, 1982).

The evolutionary, ecological life-history approach adopted here, with its focus on the phenotype and its development, suggests that to measure fitness one must assess offspring "quality" as well as quantity. While relative differences in individuals' numeric reproductive success are the *ultimate* driving force behind evolution, greater attention to the nature and development of "quality" is indicated because "quality" is a property of the phenotype, the locus of selection, and because of the K-selected evolution of low reproductive rates in the higher primates and hominids.

The evolution of "quality"

As an adaptation to the higher levels of intraspecific competition that characterize populations at carrying capacity in relatively constant or predictable environments, the essence of K-selection is to ensure reproduction at replacement levels. Since there is no advantage in reproducing at high levels, for this would intensify already high levels of intraspecific competition, K-selection works to lower reproductive rate and increase efficiency in the exploitation of already limited resources. Thus, a valid quick conceptualization of K-selection is "fewer and better" offspring. The usual interpretation of "better" is efficiency in resource exploitation, but another, more encompassing view is "more adaptable". By "adaptable" I mean the capacity of an organism to make a successful response to perturbations in its physical and social environments such that the *next time* it encounters that same perturbation, or one sufficiently similar, it can respond with less cost (see Slobodkin and Rapoport, 1974; and Chisholm, 1983 for discussions of adaptation and adaptability). Cost is generally reckoned as time, energy, risk and resources expended in responding to socioecological stressors, but in the context of our discussion of life-history theory we are more concerned with how these costs of responding to stressors may affect the developing organism's survival, optimal growth and development and learning, and ultimate reproductive value. From the perspective of life-history theory, one component of adaptability can be seen as "developmental environmental tracking", and another as "buffering" against the developmental environment. The result in either case, is an increased probability that the organism will arrive at the optimum phenotype ("quality") in a variety of developmental environments.

Perhaps the major avenue whereby K-selection achieves greater adaptability (more finely-tuned and efficient developmental environmental tracking and buffering systems) is through selection for prolonged development. In turn (according to theory, e.g. King and Wilson, 1975; Gould, 1977; Bonner, 1982), prolonged development (and/or delayed sexual maturation) seems to be achieved through selection for regulator genes which delay the expression of structural genes coding for the production of endocrine substances (like growth hormone and nerve growth factor) that are the proximate determinants of the timing of onset, the rate, and the period of growth processes. In K-selective environments prolonged development tends to increase adaptability – and reproductive value – at least in part because with longer periods of infant and juvenile dependency individuals are given more time to learn about the more competitive and complex environments that were the ultimate, evolutionary source of their prolonged development in the first place.

Such prolonged development not only provides *more time* for learning and practice, it may also produce organisms *better able* to learn. By retarding the rate of somatic development (and/or extending its period) neural structures may be given more time to grow and differentiate – and to be *affected by* the developmental environment. For example, consider Changeux's (1985) notion of "epigenesis by selective stabilization". He argues that while the basic architecture and adult number of neurons seem genetically determined and are laid down before birth, "phenotypic variability" is nonetheless *inherent* in the process of neural development. He argues that neural development proceeds by the laying down of redundant and variable synaptic typologies which provide the raw material for epigenesis – the raw material for "neural selection" – to favor the preservation or stabilization of those synapses that have *functional significance* in a particular environment.

Changeux also notes that there is a progressive *decrease* in the determining effect of the genotype on the neural phenotype from invertebrates to vertebrates, from lower vertebrates to higher, and from non-human primates to humans because of the *indeterminancy* inherent in neural developmental processes:

This phenotypic variability is intrinsic. It is the result of the precise "history" of cell division and migration, of the wandering of the growth cone and its fission, or regressive processes and selective stabilization, which cannot be exactly the same from one individual to another even if they are genetically identical. (1985, p. 247)

This greater capacity of the hominid brain *to be affected by experience* depends on both brain size and developmental timing. With *more* developing neurons, any inherent "slippage" between genotype and phenotype may be magnified by virtue of the facilitating or inhibiting effect one developing neuron may have on another. By prolonging the period of development more *time* is provided for these effects to occur. The result might often be increased adaptability, for epigenesis by selective stabilization constitutes a kind of developmental environmental tracking whereby at least some components of the phenotype are determined more *immediately* by its developmental environment than by its genotype.

Even more, the effects of the developmental environment on the organism's adult behavioral phenotype are not likely to be random with respect to socioecological or demographic factors, and the organism is not a mere receptacle for these effects. Instead, many organisms may actually *seek* the kinds of experiences and environments that have fitness-promoting long-term consequences. Fagen (1977, 1982), for example, argues that while animal play has few net immediate benefits on fitness (because of the way play diverts energy away from growth and often exposes the animal to risk) it may have considerable long-term fitness benefits. The long-term benefits come from the enhanced behavioral flexibility made possible through the power of play to increase neural interconnectivity. By altering the relationship of the young animal to its developmental environment, play effectively makes that environment an "enriched" one. Because of the long-term neural effects of development in an enriched environment, animals that play as juveniles may come to show more adaptability through a greater tendency to explore, to switch rapidly between different behavior patterns, and to reverse previous learning and engage in new learning.

Fagen suggests that higher-order taxonomic differences in play are due to differences in survivorship. In energetically inefficient species, where there is an adaptive premium on rapid attainment of adult body size (more common in high metabolism, r-selected, small animals) the costs of diverting energy from growth to play selects against the motivation to play. Among larger, more energetically efficient K-selected animals, however (with lower metabolic rates and typically higher survivorship), the costs of diverting energy from growth to play are lower. And with higher survivorship, especially among juveniles, and longer lives, K-selected animals are more likely to live enough to enjoy the long-term benefits of play.

On the other hand, Fagen argues, lower-order differences in play are due to differences in the availability and efficient utilization of energy. Play is thus facultatively suppressed in animals otherwise expected to show it when current conditions predict a low probability of living long enough to reap its benefits (e.g. disease, starvation, a particularly harsh environment, or even a stressful social environment). Horn and Rubenstein (1984) make a similar point when they argue that "behavioral decisions about life history" (i.e. phenotypic plasticity in behavior) should occur most frequently in large animals with low reproductive output who are in good condition with large amounts of stored nutrients. This is because *being* in good condition and *having* large amounts of stored nutrients are reasonable predictors of future survival – and thus the capacity to benefit from the long-term benefits of play (the neural effects of development in an "enriched" environment).

Fagen's work provides us with an example of how we can better understand the causes of development by examining the socioecological factors affecting the fitness costs and benefits associated with different developmental patterns. With respect to play, his conclusion is that when the environment varies so as to produce a *succession* of novel selective forces, play should evolve to have a general effect on behavioral flexibility. One environment in which a succession of novel selective forces is virtually guaranteed is a social environment in which individuals learn, and are surrounded by others who also learn, and who continuously generate novel behavioral strategies. This is a good characterization of the social environment of hominid evolution, at least since the end of the Pleistocene when our anatomically modern ancestors spread over the globe and adjusted their subsistence activities, work loads, sexual divison of labor, economies, social organizations, and especially their patterns of mating and reproductive effort to suit local conditions. Because of increased parental investment, especially male parental investment, hominid evolution was probably also characterized by increased juvenile survivorship (Lancaster and Lancaster, 1983), which would further increase the payoff from phenotypic plasticity in behavioral development.

Phenotypic plasticity and canalization

Phenotypic plasticity and canalization are old notions, and frequently clouded with conceptual and terminological confusion. Lately, however, evolutionary ecologists concerned with the development of the phenotype are recognizing their theoretical value and beginning to clarify their conceptual bases. This is because both are seen as processes whereby the developmental effect of the genotype on the phenotype is made less direct or immediate: phenotypic plasticity deals with developmental environmental tracking and canalization deals with buffering against the developmental environment.

Stearns (1982) has provided a summary of the most recent refinements of the concepts of phenotypic plasticity and canalization and their relevance for life-history theory. He defines phenotypic plasticity as the capacity of a single genotype to produce a wide range of environment-dependent phenotypes. Phenotypic plasticity provides an adaptive advantage when the fitness of the phenotype is more immediately determined by its developmental environment than by its genotype, which might otherwise be suboptimal. That is, through such environmental effects as maintenance, facilitation and induction (Gottlieb, 1976) the phenotype achieves greater fitness than it would without such a capacity to be affected by the environment. The range of phenotypes permitted by a genotype with the capacity for phenotypic plasticity may be wide, but the adaptive significance of phenotypic plasticity comes not from a wide reaction norm, per se, but from the phenotype's ability to specifically *track* its developmental environment, to be affected by it in specific ways that promote adaptability (or "quality") – the major component of fitness in highly K-selected organisms. Canalization, on the other hand, is the capacity of a wide range of genotypes to produce the same phenotype, which may be the optimum. Canalization provides a fitness advantage when only one or a narrow range of phenotypes results in maximal reproductive success. If phenotypic tracking is a sort of developmental environmental tracking, canalization is a buffering against some aspect of the developmental environment – or against mutations and recombinations which might also endanger development of the optimum phenotype.

Stearns discusses two sorts of phenotypic plasticity that seem relevant to human development. Continuous phenotypic plasticity describes the type of developmental environmental tracking in which the phenotypic response to the environment (i.e. the effect of the environment on the phenotype) is scaled to some feature of the environment with a continuous distribution. Within limits, for example, many organisms grow larger as a direct function of food availability. Discrete plasticity, on the other hand, describes the type of developmental environmental tracking in which the phenotypic response of the organism is not scaled to some feature of the environment, but instead is *switched* or *triggered* by the environment onto one of two (or some small number of) developmental pathways. In discrete plasticity there are few if any intermediate phenotypes, and all individuals show one or the other of only two (or some small number of) discrete phenotypes. The environmental sex determination of some fish and turtles are good examples (e.g. Charnov, 1982).

Continuous phenotypic plasticity is generally expected when socioecological conditions vary so as to present the individual with a succession of novel environments (i.e. the environment is predictably unpredictable). Under these conditions, individuals with a greater capacity for continuous plasticity are more successful in responding to novel conditions. A succession of novel environments is more likely to be encountered at the extreme K-selected end of the continuum, where there is likely to be the most crowding, the most intense sociality, and the most competition. When this happens, the adaptive value of one individual's behavioral strategy depends in great measure on the strategies adopted by all the others (e.g. Maynard Smith, 1976; Parker, 1984; Dawkins, 1980). Continuous plasticity enables individuals to generate conditional strategies in which their own behavior is scaled to that of others (Horn and Rubenstein, 1984). As mentioned, continuous plasticity is also expected more frequently in individuals in good condition – the strongest, the healthiest, and those with the most stored resources. This is because such individuals are in a better position to *use* their resources – to allocate them among the costs associated with the larger number of behavioral options that they make possible; we do not expect as much continuous phenotypic plasticity in individuals who cannot afford the price of such plasticity (see Fagen, 1982; and Johnston, 1982 for discussions of the costs and benefits of phenotypic plasticity).

On the other hand, Stearns suggests that discrete plasticity – the existence of developmental switches – is primarily associated with three conditions: first, when environmental changes are discontinuous rather than continuous (or are at least perceived that way because of threshold effects in the organism's sensory physiology); second, when intermediate or graded phenotypes could not occur or function because of prior developmental constraints; and third, when environments vary more *between* generations than within them. Fagen (1982) refines this argument by adding that such switches should be especially common when the environment fluctuates unpredictably over a *predictable* set of values. Under these circumstances development should be sufficiently plastic to match the phenotype to the predictable value, but since these predictable values are essentially invariant while it is their *timing* that is unknown, the appropriate phenotype should be "rigidly" dependent on any reliable cue about the approaching environmental change.

Canalization is expected under two general sets of conditions. First, it is expected when the environment is highly stable. This is because the capacity for phenotypic plasticity entails certain costs (e.g. "placing the phenotype at the mercy of the environment" (Johnston, 1982)), and in unchanging environments there is simply no benefit to be accrued from such plasticity. Second, however, even in environments that vary, either predictably, or especially, unpredictably, those traits that are most essential for survival, optimal growth and development, and ultimate reproductive success should be most canalized, i.e. buffered from environmental or genetic perturbation.

Life–history theory in anthropology

A perennial issue in developmental psychology is that of whether, and under what conditions, early experience can affect later behavior. While there is no shortage of claims to have shown lasting effects from this or that early experience, the fact remains that too frequently such claims cannot be replicated – and even when they are, we are too often left wondering how and why. The upshot is the growing feeling that simple maturation may account for more in development than we thought, that early experience is likely to affect later behavior only under the most stable of environmental conditions, or that early experiences are mostly irrelevant to later behavior (e.g. Bateson, 1976, 1982; Chisholm, 1983; Chisholm and Heath, 1987; Clarke and Clarke, 1976; Dunn, 1976; Kagan, 1980, 1981; Kagan *et al.*, 1980, Lerner, 1984; Sameroff, 1975; Sameroff and Chandler, 1975). There are, however, two ways that life-history theory can help us think about these issues and the causes of human development in general. First, life-history theory, through the concepts of phenotypic plasticity and canalization, provides an evolutionary ecological rationale for beginning to understand why some early experiences have long-lasting effects on the phenotype and others do not. Second, looking at human development as the development of reproductive strategies may help us appreciate which early experiences are important under different socioecological conditions.

Canalization in Navajo mother–infant attachment

The standard example of canalization in human development is that of "catch-up growth" after episodes of illness or undernutrition (Prader, Tanner, and von Harnack, 1963). The possibility of "behavioral catch-up growth" as an analogy to physical "catch-up growth" has appealed to many, but the examples offered have been mostly abstract ones. Indeed, Bateson (1976) argued that:

It may not yet be possible to give a clear instance where different developmental mechanisms generate the same behavioral end-product. Nevertheless, this could well be an area where good ones will emerge once we start to look for them. (1976, p. 410).

It was this kind of thinking that prompted me to look for evidence of canalization in my own research on use of the cradleboard among the Navajo (the cradleboard is a wooden board onto which swaddled infants are strapped several hours each day for most of the first year; see Chisholm, 1983; Chisholm and Richards, 1978 for details of theory, methods and results). Observing that the cradleboard disrupted Navajo mother–infant interaction in ways that attachment theory suggested should lead to insecure attachment, but finding no evidence of anxious or insecure attachment, I looked for reasons why these predicted effects did not occur.

In the end three, not mutually exclusive, possibilities emerged. Although the first is obvious, its very obviousness suggests that the concept of endogenous developmental rules has more relevance than may often be appreciated. My data showed clearly that while the cradleboard lowers infants' level of arousal and reactivity, they nonetheless reliably cry *in order* to be released from the cradleboard. This control by the infant is probably sufficient by itself for explaining why the cradleboard is not used enough (however much that may be) for it to have any long-term effects.

Second, it was apparent that at the end of each cradleboard session, after the infant had cried to be released, there was a highly predictable and particularly intense two or three minutes of affectionate interaction with mother. It could not be determined that these bouts of intense sociality were initiated or maintained solely by the infant, but the regularity with which they followed release from the cradleboard suggested that they might function as a very rapid behavioral equivalent of physical "catch-up growth".

Third, in what might be a specific example of an alternate developmental pathway, it was apparent that use of the cradleboard actually *increased* mother–infant proximity: infants on the cradleboard were more likely to be within arm's reach of mother than infants not on the cradleboard. It could not be demonstrated, but I suspect that the infants were ultimately responsible for this increased proximity: when on the cradleboard, infants are quieter when kept closer to their mothers, and they may thus teach their mothers to keep them close by. Although mother–infant interaction was less intense and less mutually reponse-contingent when the child was on the cradleboard, longer periods in proximity to mother may offer the cradleboard child an acceptable alternate pathway to the end-product of secure attachment. It remains an empirical question for attachment research, but it seems that under some circumstances secure attachment might be achieved as well through proximity as through the mutual reponsiveness that is presently stressed so much.

The fact that use of the cradleboard apparently has no lasting impact on mother–infant interaction does not suggest that attachment theory is somehow impoverished. Instead, it may imply the canalization of behavioral development so that secure attachment may be reached by some alternate route. I suspect that the Navajo child is buffered against the observed and hypothetically negative immediate effects of the cradleboard by his or her own responses to these immediate effects: when he tires of the cradleboard, he cries to be released; after he is released, he shows his pleasure at being released by engaging his mother in an intense bout of affectionate interaction – or at least helps to maintain such interactions by reliably showing his pleasure in his mother's suddenly renewed interest in him. Finally, he may increase his proximity to mother when he is on the cradleboard by protesting when she places him too far away from her. The buffering against the immediate and hypothetically negative effects of the cradleboard thus seem to be in the child, as if he were administering his own early – or immediate – intervention program.

While we cannot assume that secure attachment is always everywhere, biologically adaptive for everyone concerned (e.g. Draper and Harpending, 1987; Dunn, 1976; Hausfater and Hardy, 1984; Scheper-Hughes, 1985), the literature on attachment in animals is so consistent on the developmental consequences of frank attachment failure that I think we can assume that the development of normal attachment will have been a likely candidate for protection by the evolution of buffering mechanisms. Attachment failure can lead to death and severely impaired cognitive and social-emotional development, and as life-history theory predicts, those traits most essential to any component of fitness are those that should be most buffered against socioecological perturbation.

Phenotypic plasticity and the Absent Father Syndrome

Another example of recent anthropological thinking about human development in life-history terms are two approaches by Draper and Harpending (1982, 1987) to phenotypic plasticity in the sensitive period learning of alternate reproductive strategies. Critical tests of hypotheses about alternate reproductive strategies are at the forefront of research in evolutionary ecology and are an integral part of life-history theory because little is known of the developmental factors which may prompt an organism to adopt one strategy over another as an adult (Partridge and Halliday, 1984). Draper and Harpending are among the first to suggest such hypotheses regarding alternate reproductive strategies in humans.

Underlying their approach is the argument that while selection may tolerate a high degree of phenotypic plasticity in many realms of human behavior, it should favor a closer tracking of the developmental environment in the realm of sexual parenting behaviors – because these are the universal, immediate determinants of reproductive success. Thus, considerations of alternate reproductive strategies in humans should focus on the behavioral ecology of adult mating and parenting effort, and how different patterns and balances thereof may affect children's behavioral development.

The essence of Draper and Harpending's (1982) first approach to alternate human reproductive strategies is that father presence or absence during the sensitive period of the child's first five years of life serves as a developmental switch affecting the child's cognitive and behavioral styles and ultimate reproductive strategy. They suggest that human children are adapted to be sensitive to (affected by) father presence or absence because father presence or absence facilitates learning the optimal reproductive strategy in socioeconomic environments where fathers are typically present or absent.

They note that in psychology a wealth of research indicates that compared to father-present ("dad") males, father-absent ("cad") males tend to denigrate females, feminity and female authority, and have an exploitative attitude towards females; they also tend to have higher levels of interpersonal aggressiveness and higher scores on tests of verbal skills than of spatial-quantitative skills. Compared to father-present ("coy") females, on the other hand, father-absent ("fast") females tend to show a precocious interest in sex, to hold denigrating views of males and masculinity, and to show a limited ability or willingness to maintain for long an exclusive sexual-emotional bond with one male. These generalized findings about the "absent father syndrome" in the Western societies studied by psychologists are robust; even more, cross-cultural research has shown essentially the same developmental consequences in other societies. Absent-father households are typically seen as social pathology in most industrial nations, but they are the norm in a number of pre-state, tribal-level societies where social, cultural and ecological factors conspire to produce a strict sexual division of labor, separate residence for even married adults, or when activities such as warfare or long-distance hunting or trading take men away from home for long periods. In such societies, anthropologists have described a high level of antagonism between the sexes, with males and females holding low opinions of each other; rigid sex-role stereotyping, with males dominant and females submissive in virtually every realm of behavior; high levels of male aggression and display, with much male bombast and posturing; and segregation of males away from women and children in most daily activities, accompanied by an aloof, rather than intimate, tone to relations between the sexes.

The co-occurrence of these traits in non-industrial father-absent societies as well as in our own society suggests that interpreting the effects of father-absence as pathology may be an oversimplification. Instead, Draper and Harpending conceive of the effects of father-presence and absence as alternate reproductive strategies which may be adaptive in their respective environments. Aggressiveness, competitive display, and special skills in the self-interested manipulation of others are developmental consequences of early father absence for males, they suggest, because in father-absent societies sexual access to females frequently seems to depend most on position in the male dominance hierarchy. Males following such a "cad" strategy should thus be more interested in *people* than things, and their higher scores on tests of verbal ability may reflect their interests and experience in predicting and manipulating the behavior of others. On the other hand, father-present boys should develop in such a way as to optimally learn those behaviors associated with the "dad" strategy; they should be more interested in *things* than people, and greater spatial-quantitative skills or interests might better enable them to understand and exploit the natural environment for harvesting resources for investment in mate and children.

Draper and Harpending interpret the lack of any differences in verbal and spatial-quantitative skills between father-present and father-absent females by noting that female fitness has always depended more on verbal and social skills. Compared to father-present girls, however, who seem to learn that the appropriate female reproductive strategy is stable pair-bonding (one aspect of which may be granting her mate a high degree of paternity confidence in exchange for his investment in her and her children), father-absent girls act as if they are learning that without any high degree of male parental investment their best reproductive strategy is to begin their reproductive careers early. This may be, or have been, adaptive, for when females are unable to predict future resources, or such resources are likely to be poor because of low levels of mate support, they may increase their lifetime reproductive success through early reproduction (e.g. Stearns, 1976).

In sum, Draper and Harpending suggest that hominid evolution has long included a wide range of physical and social environments, and that our geographical range, our intelligence, and our highly variable economic and subsistence systems have long presented our developing young with an evolutionarily unpredictable variety of developmental tasks. Even more, with the evolution of non-genetic means of adaptation through culture, the rate of change in these developmental tasks increased by orders of magnitude. Under these conditions of "unstable diversity", continuous phenotypic plasticity is most expected. But if adult reproductive strategies have an impact on reproductive success, if conditions early in life are reasonable indicators of conditions later in life, and if the behaviors which constitute adult reproductive strategies become more efficient with practice and preparation, then a favored adaptation might be to embark early on developmental pathways which guide learning and practice in the most relevant way.

Draper and Harpending do not use the concepts of continuous

or discrete plasticity, but in talking about "developmental switches" it is clear they are thinking about discrete plasticity. Their model, however, does not require discrete plasticity, and it seems likely that there are a wide range of intermediate phenotypes between those they discuss for each sex. Further, our initial understanding of the comparative socioecology of continuous and discrete plasticity suggests that continuous plasticity is more likely in the development of reproductive strategies in the context of hominid evolution. The point remains, however, that Draper and Harpending have begun to make sense of otherwise disparate data in anthropology, evolutionary ecology, and cognitive and social-emotional development in psychology by using a life-history perspective.

Peer care, parent care and phenotypic plasticity

Applying this perspective to another set of consistent findings from cross-cultural research on early socialization, Draper and Harpending (1987) have also suggested that human mating and parenting behavior may be affected by the developmental consequences of peer care (extensive socialization by other children) and parent care (primary socialization by parents). They focus on the differences in developmental outcome between peer care and parent care of children older than 18 months for three reasons: (1) The repeated finding that early differences in patterns of cosocialization have a large impact on child behavior (e.g. Whiting, 1980; Whiting and Whiting, 1975; Chisholm, 1981, 1983; Munroe and Munroe, 1971; Ember, 1973); (2) the parent-care system predominates in modern hunting and gathering societies and in the industrial world, while the peer-care system is most frequently found in middle-range, tribal societies practicing subsistence food production; and (3) fertility seems to be lower in the parent-care, hunting and gathering and industrial societies and higher in the peer-care, tribal societies between these extremes of sociocultural complexity. Employing the approach of life-history theory, Draper and Harpending thus ask the question: are there adaptive reasons for this pattern of association between the child's early learning environment, subsistence type, and fertility? More specifically, they ask: What are the stimuli whereby developing children assess the quality of the environment in which they will act out their reproductive strategies, keeping fertility either high or low?

They suggest that at about eighteen months, when a child is capable of significant independent locomotion, he or she is faced with a developmental switch-point. Citing considerable psychological, biological, and cross-cultural research, they argue that at about this time the child is maximally disposed to match its behavior to that of its parents, but that it is also at this time that parents in peer-caretaking societies begin to place a child in social settings including only children. In peer-caretaking societies, after the age of significant independent locomotion, children typically spend the greater part of each day in a "child gang" consisting of children of both sexes up to puberty. By contrast, in parent-care societies parents keep their children closer to home, under more constant attention. In peer-care societies, subsistence activities frequently involve more work, especially for children, than they do in hunting-gathering and industrial societies, and children are frequently separated from adults by the work they do. (Draper and Harpending stress in this work that while they talk about "developmental switches" and peer- and parent-care reproductive strategies, they do so for the sake of convenience, and that there is undoubtedly a continuum between the extreme types discussed).

Spending most of their waking hours in the company of different numbers of others of quite different age/sex/kinship identities, the peer-care and parent-care children learn different things, which may, Draper and Harpending suggest, be adaptive in the different environments. Spending most of his time with the "child gang," the peer-care child learns that the best way to get what he wants (i.e. resources) is to dominate others, try to manipulate them, cajole them, or bargain with them. He also learns that "friendship, approval, information, food, and protection can be had from different peers, depending on how many there are, how variable they are, and how good the child's access to individuals in his peer network is" (Draper and Harpending, 1987, p. 223); he thus learns that resources are synonymous with his social network. On the other hand, spending more of his time with powerful adults to whom he is powerfully attached, the parent-care child learns that the best way to get what he wants is to be obedient and attentive to his parents and to emulate their behavior as best he can. He also learns that resources come primarily from only two people and are highly contingent both on his parents' efforts and his own maturation and capacities to elicit parental investment; he thus learns that resources are "scarce".

In sum, Draper and Harpending suggest that the developmental effects of these different learning environments may affect mating and parenting behavior in ways that are adaptive under the socioecological conditions that caused them in the first place. They offer the possibility that parent-reared children grow up to see that resources are scarce and require considerable effort, and that adults like this may view their own reproduction and parenting behaviors in the same light. This leads to a conservative, more "K-like" reproductive strategy, and parent-care societies thus show lower fertility than peer-care societies. On the other hand, children who spend more time in "child gangs" grow up to see that resources are not so scarce, or at least that they are more contingent on their "social skills," in the sense of being able to secure and maintain access to resources by establishing the "right" social, economic, and political connections with others (including advantageous marriages, which frequently produce many children who are valued both as laborers and as a sort of "social money in the bank", to be used for paying and incurring social, economic, and political debts with others). This leads to a less conservative, more "r-like" sort of reproductive strategy. Lancaster and Lancaster (1987) offer a similar interpretation, but focus on resource predictability instead of peer- and parent-care *per se*. They suggest that in parent-care societies resources are seen as limited but predictable, and that reproductive output is scaled accordingly. In peer-care societies, however, resources are seen as less predictable because they are seen as less contingent on one's own efforts and more on the social milieu; with resources perceived as unpredictable, the optimum reproductive strategy may be the "r-strategy" of maximum reproductive effort in hopes that some will survive.

Conclusion: some implications of life–history theory for models of human development

Perhaps the most general contribution of life–history theory will be the way it helps us generate ecologically and evolutionarily relevant developmental models. In particular, the notions of phenotypic plasticity and canalization may help us better predict which early experiences will affect later behavior, why, and under what circumstances. For example, we may paradoxically fail to reliably demonstrate that some early experiences affect later behavior precisely because that class of early experience (e.g. disrupted mother–infant interaction) frequently *did* have deleterious long-term developmental consequences during hominid evolution. If we take seriously the idea that developmental processes affect fitness, that they are exposed to natural selection and evolve, then we should also consider the possibility that selection has favored children who were affected by a *threat* to their fitness in such a manner that their response to this threat constituted a first step on an alternate developmental pathway to some adaptive end-point. We might expect, in other words, that mechanisms for the canalization of development will be found when some socioecological perturbation poses a threat to a developing child's survival or optimal preparation for reproduction – especially if the child *can* respond successfully to the threat (as the behavior of Navajo children seems to circumvent the potentially deleterious long-term effects of cradleboard use). This approach suggests that when we study the effects of early experience on later behavior we should look for the effects *immediately* after the early experience instead of after some arbitrary or convenient interval. This sort of research design would enable us to at least rule out the possibility that the immediate response of the child functions to actually *prevent* long-term effects, which might be deleterious.

On the other hand, when our data suggest that continuity in behavioral development depends on stability and continuity in the developmental environment, this is what we might expect from phenotypic plasticity. The view of phenotypic plasticity as a sort of developmental environmental tracking suggests that when certain features of the developmental environment show continuity through time, so too will those aspects of behavior that are tracking the environment; environmental continuity implies continuity in the conditions and opportunity for relevant learning and practice.

But which features of the environment should a developing child track? A life–history perspective suggests that K-selected and neotenous organisms will be most sensitive to those aspects of the developmental environments that have most consistently enhanced their ancestors' survival and optimal preparation for reproduction.

Thus, in the realm of alternate reproductive strategies at least, we might expect children to track most closely (to be most affected by) those features of their developmental environments that have provided the most reliable cues about the conditions of adult reproduction during hominid evolution. Primary among these indicators are likely to be population size and density; the quality, availability and predictability of resources; and the sex ratio and availability of potential mates. And as Draper and Harpending suggested, because sex itself is the universal and most immediate determinant of an individual's reproductive success, children may also be expected to pay particularly close attention to the mating and parenting efforts they observe in adults around them – because these adults have already committed themselves to a reproductive strategy in socioecological conditions usually similar to those the children will also face in a few years. We may thus have an evolutionary rationale for some of the most consistent findings in child development research: that differences in socioeconomic status and number and sex/age/kinship identities of others with whom the child is socialized seem to make the most reliable differences in developmental outcome.

The concept of phenotypic plasticity in no way implies that a developmental environment necessarily simply "triggers" the expression of genetic material "programming" an alternate developmental pathway. While this may in fact on occasion happen, the kinds of phenotypic plasticity discussed here deal with two (discrete phenotypic plasticity) or more (continuous phenotypic plasticity) alternate perceptual-cognitive mechanisms that are equipotential early in development but which are differentially entrained by alternate developmental environments. These alternate perceptual-cognitive mechanisms exist as a result of natural selection for the capacity to be reliably affected by those aspects of the developmental environment that have varied in the most consistent way and have had the most consistent net positive impact on "quality" (adaptability or reproductive value). They constitute a differential "sensitivity" to those aspects of the developmental environment that provide the most reliable cues about future conditions and affect behavioral development by conditionally altering the child's perceptual and cognitive predispositions and motivation to learn and practice relevant behavioral patterns (as K-selected animals in good condition, for example, are motivated to play).

Life–history theory suggests that K-selection and neoteny may frequently operate to produce "better" offspring through selection for ontogenetic mechanisms that *enable* adaptive environmental influences on the phenotype (which may themselves feed back to the genotype through genetic assimilation (Waddington, 1953) and the Baldwin Effect (Baldwin, 1896). Such processes, as Waddington saw, depend on an "essential indeterminancy" (1968, p. 364) in the ontogenetic effect of the genotype on the phenotype. This view of natural selection for the genetic capacity to be adaptively affected by the developmental environment is also the essence of Slobodkin and Rapoport's "optimal strategy of evolution":

It is not necessary for the genotype to contain a complete, detailed set of directions for the development of a particular feature if the environment itself can contribute information to the developing organism. (1974, p. 198)

Life–history theory generally, and the concepts of phenotypic plasticity and canalization more particularly, indicate that our developmental models should include considerations of what kinds of environmental information will most often have an impact on reproductive value.

Acknowledgments

My thanks to V. K. Burbank and Robert Hinde for their insightful comments on an earlier version of this chapter.

References

Baldwin, J. M. (1896), "A new factor in evolution," *American Naturalist*, 30, 441–51.
Bateson, P. P. G. (1976), "Rules and reciprocity in behavioural development," in Bateson, P. P. G. and Hinde, R. A. (eds), *Growing Points in Ethology*, Cambridge: Cambridge University Press.
Bateson, P. P. G. (1981), "Control of sensitivity to the environment during development," in Immelmann, K., Barlow, G. W., Petrinovich, L. and Main, M. (eds), *Behavioural Development*. Cambridge:

Cambridge University Press.

Bateson, P. P. G. (1982), "Behavioural development and evolutionary processes," in King's College Sociobiology Study Groups (eds), Cambridge, *Current Problems in Sociobiology,* Cambridge: Cambridge University Press, pp. 133–51.

Bonner, J. T. (1980), *The Evolution of Culture in Animals,* Princeton: Princeton University Press.

Bonner, J. T. (ed.) (1982), *Evolution and Development,* Dahlem Konferenzen, New York: Springer-Verlag.

Changeux, J-P. (1985), *Neuronal Man,* New York: Pantheon.

Charnov, E. (1982), *The Theory of Sex Allocation,* Princeton: Princeton University Press.

Chisholm, J. S. (1981), "Residence patterns and the environment of mother-infant interaction among the Navajo," in Field, T., Sostek, A., Vietze, P. and Leiderman, P. H. (eds), *Culture and Early Interactions,* Hillsdale, N.J.: Erlbaum.

Chisholm, J. S. (1983), *Navajo Infancy: An Ethological Study of Child Development,* New York: Aldine Publishing Company.

Chisholm, J. S. and Richards, M. P. M. (1978), "Swaddling, cradleboards, and the development of children," *Early Human Development,* 2 (3), 255–75.

Chisholm, J. S. and Heath G. D. (1987), "Evolution and pregnancy: a biosocial view of prenatal influences," in Super, C. and Harkness, S. (eds), *The Role of Culture in Developmental Disorder,* Orlando, Fla.: Academic Press.

Clarke, A. M. and Clarke, A. D. B. (eds) (1976), *Early Experience: Myth and Evidence,* New York: Free Press.

Daly, M. and Wilson, M. (1983), *Sex, Evolution, and Behavior,* 2nd ed., Boston: Willard Grant Press.

Dawkins, R. (1980), "Good strategy or evolutionarily stable strategy?" in Barlow, G. W. and Silverberg, J. (eds), *Sociobiology: Beyond Nature/Nurture?,* Boulder, Colorado: Westview Press, for the American Association for the Advancement of Science.

Draper, P. and Harpending, H. (1982), "Father absence and reproductive strategy: an evolutionary perspective," *Journal of Anthropological Research,* 38, 225–73.

Draper, P. and Harpending, H. (1987), "Parent investment and the child's environment," in Lancaster, J. B., Rossi, A. S., Altmann, J. and Sherrod, L. R. (eds), *Parenting Across the Life Span: Biosocial Dimensions,* New York: Aldine Publishing Company.

Dunn, J. (1976), "How far do early experiences in mother–infant relations affect later development?" in Bateson, P. P. G. and Hinde, R. A. (eds), *Growing Points in Ethology,* Cambridge: Cambridge University Press.

Ember, C. (1973), "Female task assignment and the social behavior of boys," *Ethos*, 1, 424–39.

Fagen, R. (1977), "Selection for optimal age-dependent schedules of play behavior," *American Naturalist*, 111, 395–414.

Fagen, R. (1982), "Evolutionary issues in the development of behavioral flexibility," in Bateson, P. P. G. and Klopfer, P. (eds), *Perspectives in Ethology, (vol. 5)*, New York: Plenum.

Gadgil, M. and Bossert, W. H. (1970), "Life historical consequences of natural selection," *American Naturalist*, 104, 1–24.

Gottlieb, G. (1976), "The roles of experience in the development of behavior and the nervous system," in Gottlieb, G. (ed.), *Neural and Behavioral Specificity: Studies on the Development of Behavior and the Nervous System, (vol. 3)*, New York: Academic Press.

Gould, S. J. (1977), *Ontogeny and Phylogeny*, Cambridge, Mass.: Harvard University Press.

Gould, S. J. (1982), "Change in developmental timing as a mechanism of macroevolution," in Bonner, J. T. (ed.), *Evolution and Development*, Dahlem Konferenzen, New York: Springer-Verlag.

Hausfater, G. and Hardy, S. B. (eds) (1984), *Infanticide: Comparative and Evolutionary Perspectives*, New York: Aldine Publishing Company.

Horn, H. S. and Rubenstein, D. J. (1984), "Behavioural adaptations and life history," in Krebs, J. R. and Davies, N. B. (eds), *Behavioural Ecology: An Evolutionary Approach*, Oxford: Blackwell Scientific Publications.

Johnston, T. D. (1982), "Selective costs and benefits in the evolution of learning," in Rosenblatt, J. S., Hinde, R. A., Beer, C. and Busnel, M-C. (eds), *Advances in the Study of Behavior, (vol. 12)*, New York: Academic Press.

Kagan, J. (1980), "Perspectives on continuity," in Brim, O. G. and Kagan, J. (eds), *Constancy and Change in Human Development*, Cambridge, Mass.: Harvard University Press.

Kagan, J. (1981), *The Second Year: The Emergence of Self-Awareness*, Cambridge, Mass.: Harvard University Press.

Kagan, J., Kearsley, R. B. and Zelazo, P. (1980), *Infancy: Its Place in Human Development*, Cambridge, Mass.: Harvard University Press.

King, M. C. and Wilson, A. C. (1975), "Evolution at two levels in humans and chimpanzees," *Science*, 188, 107–16.

Krebs, J. R. and Davies, N. B. (eds) (1984), *Behavioral Ecology: An Evolutionary Approach*, Sunderland, Mass.: Sinauer.

Lack, D. (1954), *The Natural Regulation of Animal Numbers*, Oxford: Oxford University Press.

Lancaster, J. B. and Lancaster, C. S. (1983), "Parental investment: the hominid adaptation,' in Ortner, D. J. (ed.), *How Humans Adapt: A Biocultural Odyssey*, Washington, D.C.: Smithsonian Institution Press.

Lancaster, J. B. and Lancaster, C. S. (1987), "The watershed: change in parental-investment and family-formation strategies in the course of human evolution," in Lancaster, J. B., Rossi, A. S., Altmann, J. and Sherrod, L. R. (eds), *Parenting Across the Lifespan: Biosocial Dimensions*, New York: Aldine Publishing Company.

Lerner, R. M. (1984), *On the Nature of Human Plasticity*, Cambridge: Cambridge University Press.

MacArthur, R. H. (1962), "Some generalized theorems of natural selection," *Proceedings of the National Academy of Sciences, 48*, 1893–97.

MacArthur, R. H. and Wilson, E. O. (1967), *The Principles of Island Biogeography*, Princeton: Princeton University Press.

Maynard Smith, J. (1976), "Evolution and the theory of games," *American Scientist, 64*, 41–5.

Munroe, R. H. and Munroe, R. L. (1971), "Household density and infant care in an East African society," *Journal of Social Psychology, 83*, 3–13.

Oyama, S. (1985), *The Ontogeny of Information*, Cambridge: Cambridge University Press.

Parker, G. A. (1984), "Evolutionarily stable strategies," in Krebs, J. R. and Davies, N. B. (eds), *Behavioural Ecology: An Evolutionary Approach*, Oxford: Blackwell Scientific Publications.

Partridge, L. and Halliday, T. (1984), "Mating patterns and mate choice," in Krebs, J. R. and Davies, N. B. (eds), *Behavioural Ecology: An Evolutionary Approach*, Oxford: Blackwell Scientific Publications.

Pianka, E. R. (1970), "On r- and K-selection," *American Naturalist, 104*, 592–7.

Plotkin, H. C. and Odling-Smee, F. J. (1979), "Learning, change, and evolution," *Advances in the Study of Behavior, 10*, 1–41.

Plotkin, H. C. and Odling-Smee, F. J. (1981), "A multiple-level model of evolution and its implications for sociobiology," *The Behavioral and Brain Sciences, 4*, 225–68.

Prader, A., Tanner, J. and von Harnack, G. (1963), "Catch-up growth following illness or starvation," *Journal of Pediatrics, 62*, 646–59.

Sameroff, A. J. (1975), "Early influences on development: fact or fancy?" *Merrill-Palmer Quarterly, 21*, 267–94.

Sameroff, A. J. and Chandler, M. J. (1975), "Reproductive risk and the continuum of caretaking casualty," in Horowitz, F. D., Hetherington, E. M., Scarr-Salapatek, S. and Siegel, G. M. (eds), *Review of Child Development Research, (vol. 4)*, Chicago: University of Chicago Press.

Scheper-Hughes, N. (1985), "Culture, scarcity, and maternal thinking: maternal detachment and infant survival in a Brazilian shantytown," *Ethos, 13*, 291–317.

Slobodkin, L. and Rapoport, A. (1974), "An optimal strategy of evolution," *Quarterly Review of Biology,* 49, 181–200.

Stearns, S. C. (1976), "Life history tactics: a review of the ideas," *Quarterly Review of Biology,* 51, 3–47.

Stearns, S. C. (1977), "The evolution of life history traits: a critique of the theories and a review of the data," *Annual Review of Ecology and Systematics,* 8, 145–71.

Stearns, S. C. (1982), "The role of development in the evolution of life histories," in Bonner, J. T. (ed.), *Evolution and Development,* Dahlem Konferenzen. New York: Springer-Verlag.

Waddington, C. H. (1953), "Genetic assimilation of an acquired character," *Evolution,* 7, 118–26.

Waddington, C. H. (1968), "The theory of evolution today," in Koestler, A. and Smythies, R. (eds), *Beyond Reductionism,* New York: Macmillan.

Waddington, C. H. (1975), *The Evolution of An Evolutionist,* Edinburgh: Edinburgh University Press.

Whiting, B. B. (1980), "Culture and social behavior: A model for the development of social behavior," *Ethos,* 8, 95–116.

Whiting, B. B. and Whiting, J. W. M. (1975), *Children of Six Cultures,* Cambridge, Mass.: Harvard University Press.

Index